Praise for

The Power of Eight

"Lynne McTaggart has demonstrated something unique in the mind-body-spirit movement: you don't need to be a Sufi master, take ayahuasca, or spend years of disciplined spiritual practice to get a taste of the miraculous. All you need is a small group with a common intention. Read this book, form your own Power of Eight group, and welcome the miracles in your own life."

—Deepak Chopra, *New York Times* bestselling author of *Quantum Healing*

"Lynne McTaggart leads the way in guiding us to an enlightened perspective on the power of consciousness."

—Marianne Williamson, *New York Times* bestselling author of *A Return to Love*

"Throw away your self-help books. As Lynne McTaggart convincingly argues in this extraordinary book, it's 'other help' that can truly transform you. As soon as you turn your thoughts from yourself to intending for someone else, you create rebound effects that heal your own life."

—Marci Shimoff, *New York Times* bestselling author of *Happy for No Reason*

"Lynne McTaggart's *The Power of Eight* is a gift to humanity, a book that truly will change your life—and in the most extraordinary and unexpected ways. This is the real secret: your life transforms when you heal for others in a group. It's at the heart of most of the world's religions, but McTaggart brings it to life, and verifies it with science."

—John Gray, *New York Times* bestselling author of *Men Are from Mars, Women Are from Venus*

"Required reading for anyone who would like proof of the power of prayer and group intention. Lynne McTaggart offers a page-turning account of her own struggles to understand the numerous extra-ordinary healing effects she has witnessed in small and large intention groups. Every church group, every book group, and every business team needs a Power of Eight group to create a vortex of healing and transformation."

—John Assaraf, *New York Times* bestselling author of *Having It All*

"Every one of us has the power to use our thoughts to heal our world. Lynne McTaggart doesn't just describe this truth, she proves it again and again in the compelling stories of people who have been transformed by her Power of Eight groups."

—Janet Bray Attwood, *New York Times* bestselling coauthor of *The Passion Test*

"*The Power of Eight* is at once a revolution and an evolution in understanding, which opens us to both proof and practice of how the powers of a group intention can transmute and transform our pathos and our pain."

—Jean Houston, PhD, chancellor, Meridian University

"This is a monumentally significant book, a once-in-a-generation work that will turn the tide in how we unleash the power of healing for each other and for the world. Drink in its meticulous and transparent scientific method, its countless uplifting stories of vivid healing breakthroughs and its breathtakingly luminous vision. Then go manifest *The Power of Eight*."

—James O'Dea, author of *Soul Awakening Practice*

"This book is a beautifully written blueprint to awaken us to the power we collectively possess to manifest miracles of healing, transformation, and peace, both in our lives and in our world. If you buy only one book this season, make it this one, for this is one of the most important and urgent messages of our time."

—*Katherine Woodward Thomas, New York Times* bestselling author of *Conscious Uncoupling*

"Lynne's research on intentional healing documents the long-recognized belief that mental qualities such as love and hope are actually the best antidotes in fending off disease; the healing resonance created feeds back and enhances the lives of each of the group's participants. *The Power of Eight* is a powerful contribution to healing our planet."

—Bruce H. Lipton, PhD, bestselling author of *The Biology of Belief*

"The perfect manual to heal your life, yourself, and others, which will teach you everything you need to know about accessing your true innate capacity—and about how healing others is also how you heal yourself."

—Joe Dispenza, DC, bestselling author of *Becoming Supernatural*

"A powerfully convincing account of the scientific study of the power of human intention. It invites all who are questioning, all who are grasping for clarity around their place in this grand universe, to find a scientific 'home' of sorts for their spiritual quest."

—UnityWorldMinistries.org

ALSO BY LYNNE McTAGGART

The Bond: Connecting Through the Space Between Us

The Intention Experiment: Using Your Thoughts to Change Your Life and the World

The Field: The Quest for the Secret Force of the Universe

What Doctors Don't Tell You: The Truth About the Dangers of Modern Medicine

Kathleen Kennedy: Her Life and Times

The Baby Brokers: The Marketing of White Babies in America

THE POWER OF EIGHT

*

Harnessing the Miraculous Energies of a Small Group to Heal Others, Your Life, and the World

Lynne McTaggart

ATRIA PAPERBACK

NEW YORK LONDON TORONTO SYDNEY NEW DELHI

ATRIA PAPERBACK
An Imprint of Simon & Schuster, Inc.
1230 Avenue of the Americas
New York, NY 10020

First Atria Paperback edition September 2018

ATRIA PAPERBACK and colophon are trademarks of Simon & Schuster, Inc.

For information about special discounts for bulk purchases, please contact Simon & Schuster Special Sales at 1-866-506-1949 or business@simonandschuster.com.

The Simon & Schuster Speakers Bureau can bring authors to your live event. For more information, or to book an event, contact the Simon & Schuster Speakers Bureau at 1-866-248-3049 or visit our website at www.simonspeakers.com.

Interior design by Kyoko Watanabe

Manufactured in the United States of America

20 19 18 17 16 15 14 13

Library of Congress Cataloging-in-Publication Data

Names: McTaggart, Lynne, author.
Title: The power of eight : harnessing the miraculous energies of a small group to heal others, your life, and the world / Lynne McTaggart.
Description: New York : Atria Books, 2017.
Identifiers: LCCN 2017008522 (print) | LCCN 2017031871 (ebook) |
ISBN 9781501115561 (E-book) | ISBN 9781501115547 (hardback) |
ISBN 9781501115554 (paperback)
Subjects: LCSH: Telepathy | Small groups. | Healing—Miscellanea. | BISAC:
BODY, MIND & SPIRIT / New Thought. | BODY, MIND & SPIRIT / Parapsychology / ESP (Clairvoyance, Precognition, Telepathy). | SCIENCE / Quantum Theory.
Classification: LCC BF1171 (ebook) | LCC BF1171 .M36 2017 (print) |
DDC 133.8/2—dc23
LC record available at https://lccn.loc.gov/2017008522

ISBN 978-1-5011-1554-7
ISBN 978-1-5011-1555-4 (pbk)
ISBN 978-1-5011-1556-1 (ebook)

For Caitlin and Kyle, and in memory of Stella,
who didn't need to see in order to believe

The miracles in fact are a retelling in small letters of the very same story which is written across the whole world in letters too large for some of us to see.

—C.S. Lewis, *God in the Dock*

CONTENTS

Prologue

For many years, I refused to write this book because I didn't believe for one moment the strange healings that were happening in my workshops, which is to say, I had a hard time handling miracles.

By "miracles" and "healings" I'm not being metaphoric; I'm referring to genuine loaves-and-fishes-type miraculous events—a series of extraordinary and untoward situations in which people were instantly healed of all sorts of physical conditions after being assembled into a small group and sent a collective healing thought. I'm talking about the kinds of miracles that defy every last belief we hold about the way we're told the world is supposed to work.

The idea of placing people into small groups of about eight started out as a crazy whim of mine during a workshop I ran in 2008, just to see what would happen if group members tried to heal one of their group through their collective thoughts. I'd billed them "Power of Eight" groups, but I may as well have called them Power of Eight Million, so potent did they turn out to be and so much did they rattle everything I thought I knew about the nature of human beings.

As a writer, I am drawn to life's great mysteries and biggest questions—the meaning of consciousness, the extrasensory experience, life after death—particularly those anomalies that upset conventional wisdom. I like to ferret out, as the psychologist William James put it, the single white crow necessary to prove that not all crows are black.

But for all my forays into the unconventional, I remain, at heart, a hard-nosed reporter, the result of my early background as an investigative journalist, and I constantly look to build an edifice of solid evidence. I am not given to arcane references to mysticism, auras, or any sloppy or inchoate uses of the terms "quantum" or "energy." In fact, there's nothing I hate more than unsubstantiated woo-woo, because it gives what I do such a bad name.

I'm not an atheist—or even an agnostic. A deeply spiritual side to me remains convinced that human beings are more than a pile of chemicals and electrical signaling. But the reason I remain drawn to the Maginot Line separating the material from the immaterial is that I rely upon bell curves and double-blind trials to underpin my faith.

My own, relatively conventional, view about the nature of reality had first taken a knocking after researching my book *The Field*. I had begun the book as an attempt to understand scientifically why homeopathy and spiritual healing work, but my research soon led me into strange new territory, a revolution in science that challenges many of the most cherished beliefs we hold about our universe and how it operates. The frontier scientists I met during the course of my research—all with impeccable credentials attached to prestigious institutions—had made astonishing discoveries about the subatomic world that seemed to overthrow the current laws of biochemistry and physics. They'd found evidence that all of reality may be connected through the Zero Point Field, an underlying quantum energy field and vast network of energy exchange. A few frontier biologists had conducted research suggesting that the primary system of communication in the body is not chemical reaction, but quantum frequency and subatomic energetic charge. They'd carried out studies showing that human consciousness is able to access information beyond the conventional bounds of time and space. In countless experiments, they'd shown that our thoughts may not be locked inside our heads but may be trespassers, capable of both traversing other people and things and even actually influencing them. Each of them had stumbled on a tiny piece of what compounded to a new science, a completely new view of the world.

Writing *The Field* hijacked me into further pursuit of the nature of this strange new view of reality. I had grown especially curious about the implication of these discoveries: that thoughts are an actual *something* with the capacity to change physical matter.

This idea continued to nag at me. A number of bestselling books had been published about the law of attraction and the power of intention—the idea that you could manifest what you most desired just by thinking about it in a focused way—but to all of this I maintained a certain incredulity, overwhelmed by a number of awkward questions. Is this a true power and exactly how all-purpose is it, I wondered. What can you do with it? Are we talking here about curing cancer or shifting a quantum particle? And to my mind, the most important question of all: What happens when lots of people are thinking the same thought at the same time? Does this magnify the effect?

From the studies I'd researched for *The Field*, it was clear that mind in some way appeared to be inextricably connected to matter and, indeed, seemed capable of altering it. But that fact, which begged many profound questions about the nature of consciousness, had been trivialized by these popular treatments into the idea that you could think yourself into great wealth.

I wanted to offer something besides manifesting a car or a diamond ring, something besides just *getting more stuff.* I had in mind a bolder enterprise. This new science seemed to change everything we thought we knew about our innate human capacities, and I wanted to test it to the limit. If we had this kind of extraordinary extended potential, it suggested that we needed to act and live differently, according to a radical new view of ourselves, as a piece of a larger whole. I wanted to examine whether this capacity was powerful enough to heal individuals, even the world. Like a twenty-first-century doubting Thomas, I was essentially looking for a way to dissect magic.

My next book, *The Intention Experiment*, intended to do this, by compiling all the credible scientific research into the power of mind over matter, but its purpose was also an invitation. Very little research had been carried out about group intention, and my plan was to fill that gap by enlisting my readers as the experimental body of group intenders in an ongoing scientific experiment. After the book's publication in 2007, I gathered together a consortium of physicists, biologists, psychologists, statisticians, and neuroscientists highly experienced in consciousness research. Periodically I would invite my internet audience, or an actual audience when I was delivering a talk or workshop somewhere, to send one designated, specific thought to affect some target in a laboratory, set up by one of the scientists I was working with, who would then calculate the results to see if our thoughts had changed anything.

Eventually this project evolved into, in effect, the world's largest global laboratory, involving several hundred thousand of my international readers from more than a hundred countries in some of the first controlled experiments on the power of mass intention to affect the physical world. Although a number of the experiments were quite rudimentary, even the simplest was carried out under rigorous scientific conditions, with painstaking protocol followed. And all but one of the experiments were conducted with one or more controls, and were also "blinded," so the scientists involved were ignorant of the target of our intentions until after the experiment was over and the results calculated.

I was far from convinced that we'd get positive results, but I was willing to give it a go. I wrote many qualifiers into *The Intention*

Experiment about how the actual outcome of the experiments didn't matter so much as simply having the willingness to explore the idea, then launched the book, kicked off the first experiment two months later, and took a deep breath.

As it turned out, the experiments did work. In fact, they *really* worked. In the thirty experiments I've run to date, twenty-six have evidenced measurable, mostly significant change, and three of the four without a positive result simply had technical issues. To put these results in perspective, almost no drug produced by the pharmaceutical industry can lay claim to that level of positive effect.

It was a year after I began the global experiments with groups of thousands that I decided to try to scale down the entire process in my workshops by creating Power of Eight groups and having them send healing intention. For me it was just another, more informal experiment, and just as foolhardy a one—until it too began to work in ways that eclipsed everything I'd imagined would happen, and people with long-standing conditions reported instant, near-miraculous healings.

The Intention Experiment caught the public imagination. Bestselling author Dan Brown even featured me and my work in one of his books, *The Lost Symbol*. But the results of the experiments themselves are only part of the story. In fact, they aren't the important part of the story.

For most of the time I was running these experiments and Power of Eight groups, I now realize, I was asking the wrong questions.

The most important questions had more to do with the process itself, and what it suggested about the nature of consciousness, our extraordinary human capacity, and the power of the collective.

The outcome of both the groups and the experiments, amazing though they were, paled in comparison to what was happening to the participants. The most powerful effect of group intention—an effect overlooked by virtually every popular book on the subject—was on the intenders themselves.

At some point I began to acknowledge that the group-intention experience itself was causing big changes in people: changing individual consciousness, removing a sense of separation and individuality, and placing members of the group in what can only be described as a state of ecstatic unity. With each experiment, no matter how large or small, whether the global experiments or the Power of Eight groups, I observed this same group dynamic, a dynamic so powerful and life-transforming that it enabled individual miracles to take place. I recorded hundreds, if not thousands, of these instantaneous miracles in participants' lives: They healed long-standing serious health conditions. They mended estranged relationships. They discovered a renewed life purpose or cast off workaday jobs in favor of a career that was more adventurous or fulfilling. A few of them even transformed right in front of me. And there was no shaman or guru present, no complex healing process involved—in fact, no previous experience necessary. The inciting instrument for all of this was simply the gathering of these people into a group.

What on earth had I done to them? At first, I didn't believe it. For years I attributed what appeared to be *rebound* effects to my imagination working overtime. As I kept telling my husband, I needed to gather more stories, carry out more experiments, amass *more hard proof.* Then I became frightened by them and sought some historical or scientific precedent.

Eventually it dawned on me that these experiments were providing me, in the most visceral way, with an immediate experience of what I previously had understood only intellectually: that the stories we tell ourselves about how our minds work are manifestly wrong. Although I'd written in *The Field* about consciousness and its effects on the big visible world, what I was witnessing surpassed even the most extravagant of these ideas. Every experiment I ran, every Power of Eight group that assembled demonstrated that thoughts are not locked inside our skulls, but find their way into other people, even into things thousands of miles away, and have the ability to change them. Thoughts weren't just things or even things that affect other things; thoughts might even have the capacity to fix whatever was broken in a human life.

This book is an attempt to make sense of all the miracles that happened in these experiments—to figure out what indeed I'd done to our participants—within the larger context of science and also esoteric and religious historical practice. It is a biography of an accident, a human endeavor that I stumbled across that appears to have ancient antecedents, even in the early Christian church. *The Power of Eight* is also about me, and what happens to someone like me, when the rules of the game—the rules by which you've lived your life—suddenly no longer apply.

The outcome of the group Intention Experiments are remarkable, but they aren't the point of this story. This story is about the miraculous power you hold inside of you to heal your own life, which gets unleashed, ironically, the moment you stop thinking about yourself.

PART I

*

Explaining a Miracle

Chapter 1

The Space of Possibilities

Communal

I was sitting at my computer with my husband, Bryan, one afternoon in late April 2008, both of us trying to figure out how we might scale down the large Intention Experiments I'd been running for the workshops we planned to hold in the United States and London the following summer.

It was a year after I'd launched the big global Intention Experiments, inviting readers from around the world to send an intention to a well-controlled target set up in the laboratory of one of the scientists who'd agreed to work with me. We'd run about four of them by that time, sending intention to simple targets like seeds and plants, and recorded some remarkably encouraging results.

Now I was trying to scale down these effects to something personal for people, something that would fit well in a weekend workshop, but I hadn't run many workshops before, and all I knew at the time was what I didn't want, which was to pretend that I could help people manifest miracles, as many similar workshops on intention were being billed. I was also preoccupied by the natural limitations of a workshop setting. The power of thoughts to affect someone's life might only become apparent over a time frame of weeks, months, or even years. How were we going to demonstrate any meaningful transformation between Friday and Sunday afternoon?

I started writing out our thoughts on a PowerPoint slide:

I typed in "Focused." I'd interviewed many masters of intention—Buddhist monks, Qigong masters, master healers—and all had reported getting into a highly energized and focused mind state.

" 'Concentrated,' " said Bryan. Perhaps a mass intention amplified this power. It certainly seemed to.

Focused

Concentrated

All the global Intention Experiments I was planning were designed to heal something on the planet, so it made sense to continue to focus on healing in the weekend workshops. We decided the workshop would try to help to heal something in our attendees' lives.

I then wrote down: "Communal."

A small group.

"Let's try putting them into little groups of eight or so and have them send a collective healing intention for someone else in the group with a health condition," I said to Bryan. Perhaps we could find out whether a tiny group had the intention horsepower of the larger groups. Where was the tipping point in terms of numbers? Did we need a critical mass of people of the magnitude of some of our larger experiments, or would just a group of eight do? We can't remember which one of us—probably Bryan, a natural headline writer—came up with it, but we christened the groups "The Power of Eight," and by the time we got to Chicago on May 17, we'd come up with a plan.

I'd started thinking about the idea of small groups after what had happened to Don Berry. A US Army veteran from Tullahoma, Tennessee, Don had written in to my Intention Experiment website forum in March 2007, offering to be our first human Intention Experiment. In 1981, he had been diagnosed with ankylosing spondylitis, and his spine was fused, making it impossible for him to move from side to side. Even his ribs seemed frozen in place, and because of his condition, his chest hadn't moved for twenty years. Over the years, he had had both hips replaced, and he was in constant pain. He had numerous X-rays and other medical test reports, he said, and so he could produce a full record of his medical history by which to measure any change.

Don's blog prompted members of my online community to set twice-weekly periods during which they would send healing intention to Don, and he in turn began to keep a diary of his condition. "While it was going on, I did start to feel better," he wrote me. "It was not an immediate healing, but my well-being was better and I was in less pain."

Don wrote me eight months later. When he'd gone for his semiannual doctor appointment with his rheumatologist, for the very first time, after his doctor asked after him, he could say that he felt absolutely fantastic, with only occasional pain. "I was (am) still fused together, but I felt I was bending more and I was *wayyyyyy* down on the pain scale," he told him. "The best I have ever remembered feeling."

The doctor then pulled out his stethoscope to listen to Don's heart and had him take a deep breath. At the end of Don's breath, as the doctor listened intently, he suddenly looked up at Don, his face incredulous, and said, "Your chest just moved!"

The doctor actually sat there with his mouth open, Don wrote me. "My chest moves!!!!!! I feel like a normal person again! I did not have a spontaneous healing, but the Intention Experiment set the wheels in motion for me to feel so much better, and it also caused me to recognize how the way I thought affected my health and even the world around me."

I thought the group effect in our Chicago workshop would be like this, some minor physical improvement caused by a placebo effect, a feel-good exercise—something akin to a massage or a facial.

I say Chicago, but we weren't anywhere near the city itself. but in Schaumburg, Illinois, one of Cook County's model villages within the Golden Corridor of northwest Illinois, so named because of the gold mine of profit from the shopping malls, industrial parks, Fortune 500 companies, and Hooters and Benihana restaurants lining Interstate 90. Motorola had placed its corporate headquarters in Schaumburg; Woodfield Mall, a stone's throw from our hotel, was the eleventh largest in the United States. We could have been anywhere in America, in one of those massive hotel complexes that sit along a highway. The Renaissance Schaumburg Convention Center Hotel had been chosen by our conference organizers largely for its location (thirteen miles from O'Hare Airport). After realizing the full economic possibilities of trading sleepy farmland for upscale suburban development, the town's urban elite had purchased a final forty-five more acres, sandwiched between the swirl of highway between 90 and Route 61, and transformed it into the elegant hotel we were presently staying at.

The evening before the conference we sat in the cavernous atrium around an electric fireplace, staring out at the small flume of water dwarfed by the giant pond separating us from a figure eight of tollways. It still felt far too early in my own process of discovery to be running this workshop, and I was worried about what was going to happen the following day. Should we be forming circles? Should everyone hold hands? Where should the person who we were aiming to heal be—in the center of the circle or as part of the circle? How long should the group hold their healing intention? And did it have to be exactly eight, or could we use any number of members in a group?

We had proceeded so cautiously in our global internet experiments, careful to avoid any human subjects in anything other than the small informal groups that had formed on the community section of my website sending healing to people like Don Berry, because we didn't know whether having thousands of people focusing their thoughts on a person would have a positive or a negative effect. For once we'd be operating with no safety net, no blinded trial or scientific method. *What if somebody got hurt?* Only one thing seemed certain, to me, even though it was only a feeling I had: the need to put the groups in a circle. Tomorrow, we told ourselves, we'd find out whether that intuition had been correct.

On Saturday, we divided our audience of a hundred into small groups of about eight, making sure that most were complete strangers. We asked someone in each group with some sort of physical or emotional condition to nominate themselves to be the object of their group's intention. They would explain their condition to the group, after which the group would form a circle, hold hands, and send healing thoughts in unison to that group member, holding the intention for ten minutes, the length of time that we'd used in our large experiments, largely because it seemed to be the maximum time that untrained people could hold a focused thought.

I instructed the audience in "Powering Up," a program that I'd created and published in *The Intention Experiment* after distilling the most common practices of intention "masters"—master healers, Qigong masters, and Buddhist monks—and synthesizing them with conditions that have worked best in mind-over-matter studies carried out in a laboratory. This technique began with a little breathing exercise, then a visualization, and an exercise in compassion to help people get into a focused, energized, heartfelt state.* I also showed them how to construct a highly specific intention, since being specific seemed to work best in laboratory studies. All the members of each group were to hold hands in a circle or place the person being

* You'll find a full summary of the Powering Up program in chapter 22, "Gathering the Eight," page 237.

targeted in the circle's center, with the other group members placing a hand on him or her like the spokes of a wheel. I had no idea which configuration was preferable, but it seemed important to maintain an unbroken physical connection between each member of the group.

"This is just another experiment of sorts," I told everyone just before we started, although what I didn't tell them was that they were on a maiden voyage and I was basically making up the route as I went along. "Any outcome you experience is acceptable." We turned on music we'd used for our large experiments and observed as the groups seemed to connect well and deeply. Before they left that evening, we asked the target people to be prepared to describe their experience and their current mental, emotional, and physical state the following morning.

"Don't invent any improvement that isn't there," I said.

Sunday morning, I asked those who'd received the intention to come forward and report on how they felt. A group of about ten people lined up at the front of the room, and we handed each of them the microphone in turn.

One of the target women, who had suffered from insomnia with night sweats, had enjoyed her first good night's sleep in years. Another woman with severe leg pain reported that her pain had increased during the session the day before but that it had diminished so much after her group intention that she had the least pain she could remember having in nine years. A chronic migraine sufferer said that when she woke up her headache was gone. Another attendee's terrible stomachache and irritable bowel syndrome had vanished. A woman who suffered from depression felt it had lifted. The stories continued in this vein for an hour.

I did not dare to look over at Bryan, I was so completely shocked. *The lame may as well have been walking.* For all that I disparage woo-woo, the biggest woo-woo was occurring right in front of me. I hoped that the results were not due simply to the power of suggestion. The group's intentions seemed to become more effective as the day wore on.

After we returned home, I did not know what to make of the entire experience. I dismissed the possibility of an instant, miraculous healing. Perhaps there was some expectation effect at work, I thought, some permission granted for the person to mobilize his own healing resources.

But throughout the next year, no matter where we were in the world, in every workshop we ran, no matter how large or small, whenever we set up our clusters of eight or so people in each group, gave them a little instruction and asked them to send intention to a group member, we were stunned witnesses to the same experience: story after story of extraordinary improvement and physical and psychic transformation.

Marekje's multiple sclerosis had made it difficult for her to walk without aids. The morning after her intention, she arrived at the workshop without her crutches.

Marcia suffered from a cataract-like opacity blocking the vision of one eye. The following day, after her group's healing intention, she claimed that her sight in that eye had been almost fully restored.

There was Heddy in Maarssen, the Netherlands, who suffered from an arthritic knee. "I couldn't bend my knee more than ninety degrees. And it was always aching, and when going up and down the stairs, it was always difficult for me," she said. "I usually had to carefully make my way down, step by step." Her Power of Eight group had placed her in the middle of the circle and sat close to her, with two of the group members placing a hand on her knee.

"At first I didn't feel anything. And then it got warm, and then my muscles started to shake, and everyone was also shaking with me. And I felt the pain going away. And a few minutes later, the pain was gone," she said.

That night Heddy was able to climb up and down the stairs easily and go to the hotel sauna. The following morning, the pain was still gone. "I got out of bed, and I was going to the shower and forgot that I had to go step by step. I just walked downstairs normally."

There was Laura's mother in Denver, who had scoliosis. After her turn as the intention target, she reported that her pain had vanished. Several months later, Laura wrote me to say that her mother's spine had altered so much that she had had to move the rearview mirror in her mother's car to accommodate her new, straightened posture.

And Paul in Miami, whose tendonitis in his left hand was so bad that he had to have a brace on it at all times, until he was the target of the Power of Eight group and stood in front of the audience the next day, showing how he could now move it perfectly.

There was Diane, who had such pain in her hip from scoliosis that she'd had to stop working out and had lost an inch in height over the past year. During the intention she'd felt intense heat and a rapid-fire, twitching response in her back. The next day, she declared, "It's like I have a new hip." And Gloria, who felt as though she were being stretched and elongated on both sides of her core during the intention for her, after which the constant pain in her lumbar spine was completely gone.

And Daniel, from Madrid, with an unusual condition that interfered with his body's ability to process vitamin D, so much so that his spine had developed a severe forward curvature, limiting his capacity to breathe. During the intention, his back felt sore, his hips felt heat, and his extremities cold. He felt an increase in pain, a sensation of his back stretching, as though it were growing. For a moment it felt like it were about to break. After the intention, Daniel said he was able to breathe normally, for the first time in years, and his posture was notably straighter.

There were hundreds, even thousands more, and each time I was standing there, watching these changes unfold right in front of me. I should have felt good about these amazing transformations, but at the time I mainly viewed them as a liability. I believed they were going to undermine my credibility in what I saw as my "real" work: the large-scale global experiments.

Which is why, for many years, I ignored what was happening. As any journalist would tell you, I buried the lede of the story. I didn't fully appreciate what people like Rosa had been trying to tell me, about the moment when her group sent her intention for her underactive thyroid: "I felt an opening in a tunnel and connecting with the universe. And if I received this I would be able to heal. It felt like I was giving and receiving healing, like I was healing myself."

Chapter 2

The First Global Experiments

A good reporter is a disrupter of the social order, her weaponry the meticulous recording of observable phenomena. You start with what is known and build on it, one fact at a time, like a scientist or a detective. Scientists can also be harbingers of inconvenient truths, as the best scientist, I'm told, likes to be proved wrong.

Both reporters and scientists begin by making certain assumptions. You construct your hypothesis, you design a way to test it, and then you sit back to see where you land. Sometimes you discover that you've been following the wrong directions and you have arrived on uncharted territory. And if you are a true explorer, you're delighted to be there because it's often when your hypothesis is wrong that you learn something radically new about the way the world works.

But how do you prove something that defies every law that you've been taught? What if your entire premise is outside the bounds of what is known or observable? *What if you were trying to locate the mathematical equation for a miracle?*

When I began our first global Intention Experiments in March 2007, fifteen months before that first workshop, we were also flying completely blind. We had no maps to draw on or precedent to follow; virtually no one had ventured into my area of inquiry—group intention. A solid body of scientific research had shown that human thoughts could change physical reality, with all manner of targets, from electrical equipment to other human beings. Dean emeritus of the Princeton University School of Engineering Robert Jahn and his colleague, psychologist Brenda Dunne, who ran the Princeton Engineering Anomalies Research (PEAR) laboratory, for instance, had spent thirty years painstakingly amassing some of the most convincing evidence about the power of directed thoughts to affect electronic machinery. Among many ingenious random event generators, or REG machines, as they came to call them, they had built special computer programs that randomly alternated the appearance of two images, say, of cowboys and Indians, each 50 percent of the time. Jahn and Dunne would place participants in front of computer screens and ask them to try to influence the machine to produce more Indians, then more cowboys. Over the course of more than two and a half million trials, Jahn and Dunne decisively demonstrated that human intention can influence these electronic devices in one or another specified direction, and their results were replicated independently by sixty-eight investigators.

The late William Braud, a psychologist and the research director of the Mind Science Foundation in San Antonio, Texas, and, later, the Institute of Transpersonal Psychology, had conducted a large body of research showing that thoughts can affect the movement of animals and have powerful effects on the autonomic nervous system (fight-or-flight mechanisms) and states of stress of human beings.

During the height of the AIDS epidemic in the 1980s, the late Dr. Elisabeth Targ devised an ingenious, highly controlled pair of studies, in which some forty remote healers across the United States were shown to improve the health and survival of terminal AIDS patients, just by sending healing thoughts, even though the healers had never met or been in contact with their patients.

Many formal and informal mass meditation groups had also reported positive results in lowering violence. The Transcendental Meditation organization, founded by the late Maharishi Mahesh Yogi, had carried out a number of studies of large groups of meditators, offering some provocative evidence that if 1 percent of the population is practicing ordinary TM meditation and the square root of 1 percent of the population is practicing TM-Sidhi, a more advanced type of meditation, violence of any sort, from murders to traffic accidents, decreases.

But almost no experiments had been conducted on the effect of lots of people sending the same thought to the same target at the same time.

Without precedent to draw upon in terms of group experiments, we were left with many imponderables. What was the best wording for an intention? Should we be specific in our intention statement or put out a general request that the target be affected in some way and let the universe decide exactly how? Should the intention senders all be in the same room together, or could each of them be in their individual homes, in front of their computer screens? If we carried out the experiment over the internet, as we'd planned to do, did our senders have to have some "live" connection with the target, like a live feed from the laboratory? Did distance matter, and would the power of intention wane the farther you were from their target? What's the optimum time to hold the thought for it to work? Will any old time do, or does the universe have to be in the mood? And was there an optimal number of participants necessary to produce a measurable effect? Like the TM studies, did we need a similar critical mass of people to have an effect?

We were going to have to test those questions, one painstaking step at a time.

Possibly the biggest question of all was who, among credible scientists, was willing to put his reputation on the line to do this work with me for free. Luckily, a number of scientists, like me, have a low-grade spirituality that infuses their lives and influences the research they want to do, and I soon discovered a willing volunteer for the earliest experiments in Dr. Gary Schwartz, a psychologist and director of the Laboratory for Advances in Consciousness and Health at the University of Arizona. Gary has impeccable credentials: Phi Beta Kappa honors and a degree from Cornell University, a PhD from Harvard, assistant professorship at Harvard, tenured professorship at Yale, and directorship of the Yale Psychophysiology Center and Yale Behavioral Medicine Clinic. Despite this impressive pedigree, by 1988, Gary felt restricted by the fusty world of East Coast academia and left it for the open spaces and open-mindedness of the University of Arizona, where, as a professor of psychology, medicine, neurology, psychiatry, and surgery, he could teach with the additional freedom to pursue essentially any research he wanted. This freedom was amplified in 2002, when Gary received a $1.8 million grant from the National Institutes of Health's National Center on Complementary and Alternative Medicine to create a Center for Frontier Medicine and Biofield Science. Gary had already carried out a wealth of experiments in energy medicine and had a full lab at his disposal, now the Laboratory for Advances in Consciousness and Health, entirely devoted to research into the nature of healing.

A stocky, ebullient man then in his early sixties with an air of constant urgency, Gary is a bustling hive of enthusiasms, which he has managed to work into his university curriculum as subjects worthy of graduate and undergraduate study.

By the time I first made contact with him, Gary's enthusiasms had veered toward the very outer boundaries of the human mind. Long before he agreed to get involved in my work he'd carried out a wealth of studies examining energy forms of healing and the nature of consciousness, including the Afterlife Experiments, a series of controlled studies, carefully designed to eliminate cheating and fraud, to test whether mediums in fact could communicate with the dead. His batch of mediums turned out to have an accuracy rate of 83 percent, producing more than eighty pieces of information about deceased relatives, from names and personal oddities to details about the nature of their deaths. Gary is generally up for almost anything, so long as it can be scientifically quantified. He was one scientist who would not be fazed by the idea of lots of people trying to fix a world problem by the power of positive thinking.

Nevertheless, like most scientists, he had an instinctive regard for caution, and he insisted that we proceed one baby step at a time in our global Intention Experiments. In science, you start at the absolute beginning with the most basic question you can ask. We'd start with "vegetable" and work our way through "mineral" and "animal" with experimental models that would be very simple initially and grow in complexity over time. First we'd use plants as targets, then perhaps water, and finally human beings.

I was crestfallen at having to start with plants. When I launched the Intention Experiment in 2007, I had big plans. I wanted to save people from burning buildings. I'd imagined big collective intentions to cure cancer, then repair the ozone layer, before moving on to ending violence in hot spots around the world.

To all these big ideas of mine, Gary would continuously quote from the opening scene of the movie *Contact*, where Ted Arroway is counseling his impulsive young daughter Ellie, who is on their ham radio and trying to get someone out there to tune in, secretly hoping that that particular someone will be from outer space.

"Small moves, Ellie," Gary said to me repeatedly, echoing Arroway. "Small moves." As he kept reminding me, we were working with an experimental design that had never been attempted before. First we had to establish that the thoughts of a group of people could have an effect—any effect. Only after we'd been able to demonstrate that could we even consider larger and more extravagant targets. We had to take it one faltering baby step at a time.

Although I appreciated his desire for rigor, any decisions continued to be a tug-of-war between my blue-sky imaginings and Gary's scientific caution. Okay, let's see if we can lower global warming, I'd say to Gary during one of our periodic brainstorming sessions over the phone.

How about we start with a leaf, he'd reply. And when we're done, we'll move on to seeds.

As Gary ultimately convinced me, targeting one of the simpler biological systems like a leaf or seeds helps to limit the endless variables present in a living thing, the myriad chemical and electrical processes occurring simultaneously every moment. Only by starting off with the simpler types of biological systems could we demonstrate that any changes were caused by the power of intention and not a vast array of other possibilities. And plants were clean, safe subjects. Using a plant or indeed anything nonhuman as our subject meant we wouldn't have to put our proposed study through a university internal review board, set up to make sure that human studies are carried out ethically, which could easily hold up the experiment for months.

The kind of scientific experiments we could carry out would normally be limited by the measuring equipment we could get our hands on. Fortunately, Gary's lab, housed in an unassuming one-story modern pink stucco block at the university, was a sprawling Aladdin's cave of sophisticated equipment capable of registering the tiniest flicker of change in a living organism. Gary had been highly influenced by the work of the late German physicist Fritz-Albert Popp, who, while investigating a cure for cancer, discovered that all living things, from algae to human beings, emit a tiny current of light. Popp gave his discovery the ponderous title of "biophoton emissions" and spent the rest of his life attempting to convince the scientific establishment that this dribble of faint light represented a living organism's primary means of communicating within itself and also the outside world. He grew to believe that this light was nothing less than the body's central conductor, responsible for coordinating millions of molecular reactions inside the body and its central means of orienting itself to its environment via a two-way communication system. By the time of his death in 2014, the German government and more than fifty scientists around the world had come to agree with him.

Popp had built and developed several photomultipliers to record this faint light, and Gary wanted to take this one stage further by using equipment to actually photograph it. He'd already managed to talk a professor of radiology into giving him access to a $100,000 digital charged-coupled device (CCD) camera system ordinarily used in astronomy and capable of photographing the faintest light from distant galaxies after realizing such a device would enable him to create digital photos of the subtle light emissions from living things and to count them, one pixel at a time. Even before we began working together, he had used forty thousand dollars of his grant money to purchase his own, cheaper version, a piece of equipment that would enable us to start from the ground floor. Measuring whether the power of thoughts has any effect in changing this tiny current of light would be far more subtle than, say, examining our effect on growth rate, and equipment this sensitive would enable us to capture every single hairbreadth of difference in the change of light emissions from any living thing, Gary assured me.

When scientists start out on a new experiment, the usual first step is to stand on the shoulders of the people who have ventured into the territory before, which is why they like to repeat what has already been demonstrated in previous experiments before investigating something new. For our trial run, we aimed to replicate the pilot study we'd carried out with Fritz Popp, which I wrote about in *The Intention Experiment*. In that experiment, sixteen experienced meditators and I had assembled in London and sent healing intention to four targets in Popp's lab in Neuss, Germany—two kinds of algae, a jade plant, and a woman—all of whom had been stressed in some way. Measurements of all the targets showed we'd had a strong effect in changing the tiny light emissions during the times we'd sent healing intention.

For that first Intention Experiment, however, we didn't have what most true scientific experiments insist on: a set of matched controls—similar subjects that, unlike the targets, are not the object of any changing agent. Our "control" had been just the half-hour periods when we were resting and not "running" intention, and also the fact that we'd not told the scientists when we were doing what. This time Gary and I would have two near-identical targets subjected to the same conditions; we would randomly pick one of them to send intention to and use the other as our control. Once again the studies would be "blinded"; the scientists would be kept "blind" about which target we'd chosen until after they'd calculated the results, so that no unconscious bias would influence the results in any way.

After kicking around a number of possibilities for the first major Intention Experiment, we eventually decided on a geranium leaf, picked off a thriving geranium plant in Gary's Arizona lab, as our first target. We enlisted the attendees of a conference run by my company on March 11, 2007, and asked them to send intention to lower the "glow" of one of the two geranium leaves we'd randomly selected, which would be constantly photographed by a webcam shown to our audience via a gigantic screen.

A major reason for Gary's initial caution in the choice of our first target had to do with the way you have to prove something scientifically. To show that something works according to standard scientific protocol, you need to demonstrate statistical significance, a mathematical demonstration that your results weren't arrived at accidentally but as a consequence of whatever it is you are testing, and to do that, you need a certain critical mass of the item you are studying. Science agrees that significance is anything less than "$p<.05$," which indicates less than a one in twenty chance that your results aren't arrived at by chance.

For our results to achieve true statistical significance, we needed more than thirty points of comparison between the two leaves, or what scientists call "data points." Managing that in even a rudimentary experiment like ours would nevertheless require a painstaking fifty-step protocol followed by Gary's young lab technician, Mark Boccuzzi. Mark would select two geranium leaves identical in size and number of light emissions and then puncture each leaf sixteen times in a 1.6 inch by 1.6 inch grid, a process that would take several hours of preparation. The plan was for Mark to place both leaves under his digital cameras, send the images to Peter, our first Intention Experiment webmaster, and then stand by, waiting for the signal that our group intention was finished, at which point he would use the CCD camera to photograph each leaf.

Originally, we planned to have our participants attempt to lower the light emissions, as we had in the Popp experiment. But the healthy light of a living organism defies common sense; the lower the level of light in the main, the healthier the organism. As March 11 approached, I began to think that our participants would naturally feel prompted to *increase* it. Right before the experiment, Gary and I decided to reverse the instruction in order to increase the leaf's light emissions. I wasn't thrilled about this experimental protocol. When you deliberately increase something's light, you are actually stressing it. So essentially our entire experiment was to be an exercise in injuring a living thing, even if it was just one leaf about to drop off a plant.

On the day of the conference we flipped a coin to decide our target; the other leaf would act as our control.

Gary and I had also decided that participants should focus their minds on the intention for ten minutes, an arbitrary time frame that we had chosen because we felt that any longer would be hard for them to sustain. An hour before the experiment was to begin, I'd started to worry that asking the audience to focus their thoughts for that amount of time might prove difficult without some sort of mind anchor. I asked my husband, Bryan, to consult with Mel Carlile, who ran the Mind-Body-Spirit bookshop for our conference, if he might be able to recommend meditation music for us to play during the experiment. "Here, try the first track, 'Choku Rei,'" said Mel, handing Bryan a CD of Jonathan Goldman's album *Reiki Chants*.

Right before we began, Gary got on the phone to wish the audience good luck. "Remember," he added, "you are making scientific history."

A giant image of our leaf appeared on the screen. I instructed the audience in the "Powering Up" techniques that I'd created and published in *The Intention Experiment*. "Make the little leaf glow and glow," I said. "Imagine it glowing in your mind's eye."

The intenders kept holding their intention while the hypnotic meditative music played in the background for ten minutes. Later I was astonished to discover that Choku Rei essentially means strengthening the power, flow, and focus of healing energy—something akin to "Powering Up." Maybe there *are* no accidents.

At the time, though, I felt ridiculous standing there on that stage. In my investigative reporter days, I had been fastidious about the standard journalistic practice of gathering at least two independent sources of evidence as the minimum requirement before I would regard something as fact. It was such a hard-and-fast rule with me that one evening in 1979, during the writing of my first book, *The Baby Brokers*, an exposé of the private adoption market, I had stayed up that entire night, poring over what I had about a fellow who'd set up a string of adoption agencies in different states and countries. His practices seemed highly dubious, and he'd even made a veiled threat during one phone interview, but I was weighed down by the knowledge that one slipup could unfairly ruin this person's life, even if, to all appearances, that person was somebody trafficking in human beings. I don't have it, I finally decided at 6 a.m. about one accusation I'd been preparing to make: *I can't confirm this as a fact.* Although my gut forcefully argued otherwise, I softened the story.

And now, here I was, all these years later leading my audience in a prayer to a leaf. Everything about this was violating my own two-sources rule. In fact, it was violating every last bit of the pertinacious, fact-gathering side of me.

After the ten-minute-intention period was completed, Mark placed both leaves in the biophoton imaging system and photographed them for two hours. The conference ended, everyone went back to their respective countries, and we waited to hear what had occurred on the other side of the world in a small laboratory in Arizona.

"You won't believe it," Gary said giddily on the phone to me several days later, after I'd revealed which leaf had been the target. "The leaf sent intention was glowing so much compared with the other leaf that it seems like the other leaf had a 'neglect effect.' "

The changes from the "glow" intention had been so pronounced that they were visible in the digital images created by the CCD camera. Numerically, the increased biophoton effect was also highly statistically significant. In fact, said Gary, all the punctured holes in the chosen leaf had been filled with light, compared with the holes in the control leaf, which were discernibly less bright.

A week later, he copied me in on an email to Mark: "WAIT UNTIL YOU SEE HOW PRETTY THE DATA ARE . . . images, graphs, and tables . . ." For our official press release about the event, his tone was more measured: "For a first experiment of this kind," he wrote, "the results could not be more encouraging."

Buoyed by this overwhelming result, we made plans for our first big online event to run on March 24. One of the cornerstones of our early hypotheses was that the experiment would only work if the audience had some sort of real-time connection with the target, and to achieve this we'd decided to have two live web-camera images of our target and our controls continuously running. At the last minute, Peter, our webmaster, cautioned us not to use webcam images, as we had in our conference experiment, because they might cause the website to freeze if we had thousands of online participants coming onto our website, as appeared likely as the date of the experiment approached. "Inherently webcasts always fail or at the least are very unpredictable," he wrote.

Mark had set up the next best possibility: two digital cameras, which would feed our website with a continually refreshed image of both leaves every fifteen seconds. Instead of live video of the leaves, we'd be showing live pictures of them to save server power.

My youngest daughter, then age ten, flipped a coin to choose leaf one or two. We'd upgraded to a huge server with extra RAM and turned off all our other websites to maximize our power. And then we waited once again, expecting to carry on making scientific history.

Instead, like thousands of others, I spent the next hour in total frustration, repeatedly trying to enter my website but unable to get past the first page. Peter's misgivings had proved correct. So many people—an estimated ten thousand—had tried to gain access to the site simultaneously that our website had crashed.

The only thing to do was to announce what had happened and promise our participants that we'd try again as soon as we could, while privately vowing not to promote these experiments so widely in the future so that the size of our audience could again overwhelm our server. Demonstrating the power of intention seemed to be the easy bit. The difficult part was setting up an online technical configuration that would allow thousands of people to see the same live target at the same time.

To avoid another cyber traffic jam we enlisted a new web team and rented a gigantic server from a company that supplied the servers for *Pop Idol*, the British equivalent of *American Idol*, with nine linked servers to handle the load. Our new webmaster, Tony Wood, from Vision with Technology, and his team, who'd handled the web needs for big companies like the *Financial Times*, were confident that they could construct something to keep the website from freezing. This time around we'd shut down the home page during the actual experiment, create an HTML page only visible to registered participants of the experiment, and move the actual experiment away from our main Intention Experiment site. But just to be on the safe side, Tony wanted to have a trial run a week before the real event.

On April 21, the day of the trial run, all but a handful of the more than seven thousand people who'd signed up for the experiment got into the site and were able to participate in our "glow" experiment. This time our object was string bean seeds. Again our study worked. As with our March experiment with the audience, we got a positive effect, although not a significant one in scientific terms. This was probably due to the limitations of the CCD camera, which enabled us to photograph only twelve seeds; a minimum requirement for statistical significance is to have at least twenty data points to compare. Although the first experiment concerned only two leaves, Mark had punctured them thirty times each to have more than enough data points, as he would be comparing light emissions coming from each puncture site. This time, with twelve seeds, we had only twelve data points for examining our light emissions. As Gary wrote me about the latest experiment, "If it was possible to image twice as many seeds, the results would have reached statistical significance."

But when we ran what we were billing as the "actual" experiment the week later on April 28, again with geranium leaves, only five hundred people managed to get through, and the results were inconclusive. In August we decided to go back to basics and repeat our first leaf experiment at a conference in Los Angeles, which replicated our initial results.

Despite this shaky start, we'd answered the biggest question of all:

Was any of it actually going to work? Although we'd had two inconclusive results when potential participants couldn't get onto our web portal, for those times that the bulk of participants did manage to gain access to an image of the target, we'd had three positive results: the March 11 experiment with my London conference audience; the April 21 seed experiment over the internet; and the repeat of our first leaf experiment at the Los Angeles conference. And we'd had strong effects in all three.

You begin by making certain assumptions, you construct your hypothesis, and you hope that a map appears. More experiments had worked than not worked, but beyond that, we didn't have much else to go on. Had the failures been due only to technical hitches? Or in the case of the April 21 experiment, because of low attendance? Although the website hadn't held up during that aborted March 24 internet experiment, a number of people who couldn't get on the site nevertheless had sent intention to a mental image of a geranium leaf, and Gary's sensitive CCD camera and equipment had picked up some sort of an effect, with a strong trend in the same direction as the conference data.

What did that mean? Was it pure coincidence? Was the lack of significance because the audience had not seen an image of the target, or because of the technical hitches, or was it the fact that the participants were scattered around the globe instead of in the same room, as they had been during our March 11 experiment in London? Did group intention work better when carried out by a focused group in the same physical space, as we'd had at the March conference? Or was it that you had to actually "see" the target in order to affect it?

Was the April 28 experiment unsuccessful because we didn't have a large enough critical mass of participants, or because of technical problems? Or could it have been, as Gary postulated, "a boredom effect"—my audience had grown tired of participating in essentially the same experiment?

At this point, we did not know the answer to any of these questions. In science, if you find something untoward, you take comfort in the idea that you can test it again, and if it replicates, you can locate the agent of change, and reestablish order, certainty, and a predictable arc of cause and effect.

There was only one thing we'd grown sure of: we'd have to abandon the idea of a live connection with our targets. I simply couldn't afford to rent giant server capacity every time we wanted to run an experiment. The scientists had always generously donated their time, but when I'd first come up with the Intention Experiment, I hadn't counted on all the technical costs. To keep the April 21 experiment holding up, it had cost us about nine thousand dollars for a half hour in server power alone, and the creation of special web pages many thousands more—far too much for me or my company to donate on a regular basis. We had to find another way to carry out these experiments and to come up with a study design that was readily replicable so that our results would have some scientific validity.

Basically I had to find the impossible: an enormous amount of server power, a way in which to carry out the experiments cheaply, and a platform that was going to hold up under the pressure of thousands of simultaneous visitors.

But it turned out that the live images didn't matter at all. When a batch of small groups began spontaneously forming on my internet site and having their own effects on individual members, I realized that all of us, from our separate computer screens, had already made the connection.

Chapter 3

Virtual Entanglement

Later, doctors would tell Daniel that he was lucky—it could have been his face. There'd been a horrific gas explosion at work, and his hands had been so badly burned that after he'd been rushed to the hospital, doctors told his wife he'd be needing skin grafts and weeks in intensive care. Feeling helpless and in a distraught state, she contacted a little intention group, which she and Daniel had created on my website.

By 2008 after those first expensive leaf and seed experiments, we'd decided to set up any future large-scale experiments on Ning, an online platform enabling people to create custom social networks. Ning offered two things we needed: hundreds of distributed servers, capable of handling unlimited bandwidth and unlimited numbers of participants simultaneously accessing our site, and most important, a free platform. But it also had a community site for our participants to join, where they could create their own small intention circles.

Daniel and a few other community members had created a little group on Ning and been experimenting in sending intention to each other. Hearing of Daniel's plight, the group now had a real-life target. They began sending him a daily intention at set times every day.

Five days later, Daniel left the hospital. He'd begun to heal weeks earlier than normal and confounded all expectations by not needing skin grafts. His doctors wanted to study him as a medical miracle. By way of comparison, an associate of Daniel's who had sustained near-identical injuries had stuck to orthodox healing methods. He remained in intensive care for another two weeks and went on to get skin grafts.

I was in front of an audience in Dallas in April 2008, elaborating on charts and graphs on a PowerPoint screen about our Intention Experiment results when Daniel raised his hand, still covered in what appeared to be a gauzy glove, to tell me his story.

"Since there were two of us with near-identical injuries, you can consider my experience as a controlled experiment," he said with a laugh.

I returned to my seeds and leaves and graphs, but I'd been blindsided. The rational part of me knew that we couldn't really compare Daniel and his buddy unless we controlled for all sorts of biological variables, but suppose he was right? Was it just the power of Daniel's belief—*his expectation of healing*—or was there amplified power in a group whose members were not in the same place but sending intention virtually?

Fact: Daniel and his colleague sustained similar injuries.

Fact: Daniel was the only one who'd had the group healing intention.

Fact: Daniel alone defied all prognoses and became, as his doctors were calling it, a "medical miracle."

With a miracle, you don't try to understand it by starting at the beginning, you start at the end with the bald fact of it, like walking into a room and discovering a dead body. You try to work your way back to the point where it veered off the path of known possibility, like a detective looking for the few telling fibers of cloth left on the sofa, any slight clue to deduce a credible cause.

You can't isolate the single agent of change; you can only try to create a favorable environment to coax it into reappearing. I got home and over that summer decided to play around a bit more with groups. The experiences of Daniel's and Don Berry's spinal improvements had sparked an idea. Perhaps we could run regular group intentions for people like Daniel and Don—which we would call our Intention of the Week. We could treat this as one more kind of informal experiment—a larger version of the Power of Eight groups that I was running in my workshops.

I invited my email community to participate in a weekly intention from our website—usually to try to heal someone with a health challenge or ease someone's financial difficulties in the wake of the 2008 financial crisis that autumn. We invited the web community to nominate an intention of the week, and we posted the person's name, condition, and photo on our website to send healing intention every Sunday at 1 p.m. eastern standard time.

Before long I was receiving dozens of requests every week: people with cancer or traumatic injury; children with brain damage or birth defects; members with impending bankruptcies or job losses; estranged families and wounded pets. The website was turning into the cyber equivalent of a weekly prayer group.

Our intentions didn't always work. We attracted many requests from patients who were weeks away from death. And it didn't always work in the Power of Eight groups I set up in the workshops. In most cases, we had no reports by doctors or other health professionals to independently verify the effects claimed by family members of our target intendee. Sometimes the effects were enormous—two readers claimed to have had spontaneous remissions of their cancer—and other times fleeting, but there were enough testimonials of extraordinary improvement for me to think that something was going on.

Brian had been left paralyzed and was still unconscious from a recent major accident, and his family sent in a request for him to be one of our targets. Right after Brian's healing intention, his mother started noticing that he was becoming more aware of his surroundings and paying closer attention with an overall increase in consciousness. He began answering questions more often than he had before and even began to initiate conversation.

Two days after the intention for him, Brian went to physical therapy, and for the first time walked sixty feet with the therapist and his walker and then another forty feet without a brace on the right leg. He also began using his right arm more, and was able to start riding a recumbent bike in therapy. He'd regained movement months earlier than his doctors had predicted. Margaret, a family friend, who'd nominated Brian as an Intention of the Week, wrote in with a progress report. Brian's family was "amazed with the increase in his progress," she said. To their mind, the group intention had triggered some form of "divine intervention."

Miracle. Amazement. Divine. Against all expectations.

The more I heard words like these and stories like Brian's, the more unsettled I became and the more tightly controlled I sought to make the large-scale global experiments I continued to run in tandem throughout 2007 and 2008. Gary and I decided to return to seeds, but this time with some application to real life: we'd try to affect their growth rate and health. We settled on barley seeds because they are both a common grain fed to livestock and a healthful grain for humans. We would be asking a question with huge practical implications: Can food grow faster and be healthier when sent good thoughts?

And this time, a few scientists had navigated a path of sorts before us, with several similar studies showing that seeds sent intention by a healer or irrigated with water held by the healer were healthier and had a faster germination rate and growth. These small studies were intriguing, but all had involved single individuals sending intention to seeds right in front of them. With this experiment we would be investigating whether we could attain the same outcome—or even a greater effect on our targets—if there were an entire group of disparate intenders and they were sending their thoughts from thousands of miles away.

For each of these experiments, Gary and his lab team prepared four trays of thirty barley seeds apiece—one target and three controls—to eliminate chance findings. This time, the best we could offer our audience, to connect with the target, was a photograph, although we were far from sure it was going to work. Mark decided that he would simply photograph the four sets of seeds with an ordinary camera and send them to me the night before the experiment.

I had plans to lecture in different parts of the world during that time, which gave us an ideal opportunity to test whether the experiment would work in many situations and without the worries of whether our website would hold up. My first port of call was Australia, a four-hour talk in front of seven hundred people at a very slick conference whose organizers had flown me there first-class and sent around my photo to the entire hotel staff so I'd be given star treatment.

The night before the first experiment, Mark sent me photographs of four batches of thirty seeds, each sitting in a little half circle in a seedling tray labeled A, B, C, or D, and I incorporated each image on a slide in my PowerPoint. During my live lecture the following day I had a member of the audience choose our target among the four sets of seeds, then I simply projected the photograph of the targeted seeds while leading the audience in the intention for the seeds to enjoy enhanced growth and health, again playing "Choku Rei," which I've played for every experiment, since the very first one at that 2007 London conference, in order to maintain consistency.

Once we'd finished, I called Mark to tell him we were done—the signal for him to plant all four trays of seeds. At the end of five days, he harvested the seedlings and measured their lengths in millimeters. I then had to patiently wait for weeks while Gary did his calculations, which had to be squeezed in between his own frantic teaching and writing schedule.

Gary had labeled this and the subsequent trials like this the "Intention Studies," but in order to eliminate the possibility that our results were a chance finding, or down to something other than our audience's intention, he also ran a completely separate series of control experiments after each Intention Experiment. Scientists often conduct control experiments that mimic the actual experiment in every regard without using any sort of changing agent in order to eliminate the possibility that any change observed in the actual experiment is caused by something other than the agent itself. With these control experiments, Mark set up the experiment in an identical fashion to the regular Intention Studies, selecting and preparing another 120 seeds into four sets and randomly choosing one set as the "intention target," but this time, there would be no actual intenders and no actual intention sent to the target. After a set period of time, as with the genuine experiments, he would plant all four sets of seeds, then harvest and measure them after five days.

If these control experiments showed little or no difference in the seed growth, compared to our actual Intention Studies, this would confirm that intention had been the only cause of the change. This experiment was to act as a second-tier control. It would also give us double the number of seeds to compare—1,440 in total—which also offered the possibility of greater statistical significance.

In the summer of 2007, we ran two more barley seed experiments—one with a small internet audience, and another one in front of my audience of about a hundred at Omega Institute, a retreat center in Rhinebeck, New York, offering courses in human potential.

After the Rhinebeck experiment, Gary analyzed the three studies. The results were intriguing. The first and second experiment had significant results, but the third experiment was off the charts. He sent me a first graph, to show the difference between the seeds sent intentions and the controls, a difference of 4 millimeters (0.15 inch), which sounds small but is large enough to be considered significant in a scientific study. He signed off with, "Exciting, yes?"

The third run with our Rhinebeck group with the smallest audience had produced the biggest results. It seemed logical that the bigger the group, the larger the effect, but apparently we didn't need a certain critical mass to affect our target. Were the specific growth instructions responsible, or was it the experience of the audience, most of whom were highly motivated and seasoned meditators, or even the retreat setting, with its possibility of a greater degree of focus than what's afforded when taking time out to send intentions during ordinary life?

As any scientist will tell you, a single experimental result is meaningless. The outcome can be pure coincidence—an artifact, as scientists call it. It's only when your study is replicated many times that you can say with any certainty that you've found a true effect. The only thing for us to do was to repeat the experiment a few times more in order to demonstrate that we were onto something real.

We carried out three more Germination Experiments: in Hilton Head, South Carolina, in front of 500 Healing Touch healers; before a workshop of 130 at an Association for Global New Thought conference in Palm Springs, California; and at a retreat workshop at the Crossings in Austin, Texas, with 120 attendees. After we'd run the sixth experiment, Gary analyzed the results formally through a number of complex analyses, comparing the growth of the targeted seeds with the non-targeted seeds in our Intention Studies; all genuine targets with control study "targets"; and the growth of all the seeds in the Intention Studies against the growth of all the seeds in the control studies. He used two types of statistics largely to compensate for the fact that some seeds didn't sprout at all and others grew wildly longer than usual.

"In a word, the results are STUNNING," Gary wrote me on March 16.

As an overall average, the seeds sent intention grew significantly higher than the controls in the Intention Studies: 2.20 inches versus 1.89 inches (56 mm versus 48 mm). In the control studies, there was no difference between target seeds and non-targeted seeds; in fact, the seeds labeled "Intention seeds" in the control studies grew 1.79 inches (45mm)—0.07 inches (2 mm) *shorter* than the non-targeted seeds, and higher than all the other seeds in each control study. Our effect in the Intention Studies was statistically significant; there was only a 0.7 percent possibility that we arrived at this result by simple chance.

To get a sense of how meaningful this result was, imagine that you're playing a game with a coin and you're trying to achieve a certain number of heads in a row. With our experiment, you'd have to toss that coin 143 times to reach the same result just by chance. Those targeted during real Intention Studies grew significantly higher than those "targeted" in the control studies, with 0.3 percent possibility that this was due to chance—like a coin tossed 333 times.

But the largest effect of all resulted when comparing the results of all plant growth in the actual Intention Experiments against all plant growth in the control experiments. On the days we sent intention, *all* the plants in the Intention Studies grew higher than had all the plants in the control experiments, with the plants sent intention the highest of all, as though there were some sort of communication between all the seeds in the Intention Studies. This effect was mind-boggling—a 10 million to one possibility that we'd arrived here by simple chance.

What did that mean? Does intention have a "scattergun effect"? Are living things affected by the energy of human thought from the entire environment, and not simply between two communicating entities? I thought of an experiment by Dutch psychologist Eduard Van Wijk, who had carried out numerous studies on the mysterious light emissions discovered by Fritz Popp. Van Wijk had placed a jar of simple algae near a healer and his patient, then measured the photon emissions of the algae during healing sessions and periods of rest. After analyzing the data, he discovered remarkable alterations in the photon count of the algae. The quality and rhythms of emissions significantly changed during the healing sessions, as though the algae were also affected by the healing intention.

Gary wrote up the results of all our barley-seed experiments, then presented them at the Society for Scientific Exploration's annual meeting in June 2008, and published the summary in the meeting's proceedings. It was the Intention Experiment's first attempt at formally establishing the validity of our data, and our conclusion was unequivocal: "Group intention can have selective effects on increasing the growth of seeds."

I quietly wrestled with the implications of this perfect little experiment. Within the neutral, guarded language of our modest paper lay some profound discoveries about the nature of consciousness. We'd repeatedly demonstrated that the human mind has the ability to move beyond time and space, connect with other minds, and act on matter at a distance. Essentially, we'd demonstrated something extraordinary and profound: that human minds have the capacity to operate non-locally.

Non-locality, also referred to, rather poetically, as "entanglement," is a strange feature of quantum particles. Once subatomic particles such as electrons or photons are in contact, they are forever influenced by each other for no apparent reason, over any time or any distance, despite the absence of physical force like a push or a kick—all the usual things that physicists understand are necessary for one thing to affect something else.

When particles are entangled, the actions of one will always influence the other, no matter how far they are separated. Once they have connected, the measurement of one subatomic particle instantaneously affects the position of the second one. The two subatomic parties continue to talk to each other, and whatever happens to one is identical to, or the opposite of, what happens to the other.

Although modern physicists readily accept that non-locality is a given feature of the quantum world, they maintain that this strange, counterintuitive property of the subatomic universe does not apply to anything bigger than an electron. Once things get to the sticks-and-stones levels of the world we live in, they claim, the universe starts behaving itself again, according to predictable, measurable, Newtonian laws. A few studies with crystals and algae have hinted at the fact that non-locality exists in the big measurable world and may even be the driving principle behind photosynthesis, but this property is still regarded as the exclusive domain of the tiny, the "spooky action at a distance," as Albert Einstein so famously put it, of the quantum world, and certainly not the domain of human consciousness.

Nevertheless, our little seed experiment had showed that we could create non-locality out in the big visible world, not only between the minds of individuals but also with a remote target. A group of people in Sydney, Australia, had affected seeds sitting at the University of Arizona's labs in Tucson eight thousand miles away, just through the power of one focused thought. And our intenders didn't even have to be in the same location; a group of people scattered around the globe produced the same effect as a group clustered in the same room. Somehow, like a pair of entangled electrons, our individual minds, all at a distance from one another, had made an invisible connection that was able to act collectively to alter a set of seeds, also at a distance.

I began to ponder the possibility that human consciousness possesses the ability to create a sort of *psychic internet*, allowing us to be in touch with everything at every moment. All we might require is focused attention to log on to it.

Chapter 4

Mental Trespassers

The Power of Eight groups could also create a psychic internet, which I first discovered during an intention for John, who had been the victim of a serious motorcycle accident. His mother attended one of our workshops shortly after the accident and reported that he had sustained a serious injury to his neck and several vertebrae of his spine. Doctors had told his mother that the injury to his spinal cord was so severe that he might even be left a quadriplegic.

That weekend John's mother asked her Power of Eight group to send a special intention to her son. Two months later she wrote me with a progress report: after our intention and follow-up intentions by his family members, her son began to move his upper body and had even been able to move his toes.

"He is experiencing an amazing recovery. He is probably about 85 percent back to normal—which doctors thought would take him six months to a year—not six weeks!"

If John's remarkable progress had anything to do with the Power of Eight group, they'd achieved it without any link to him: no live connection or photo of him, no knowledge of him or his whereabouts, no connection with him whatsoever except his mother and her thoughts about him.

I began to consider that a group "prayer" circle could create an enhanced healing environment, and the group has the ability to make some sort of invisible connection, the same sort of extraordinary connection we'd been witnessing with the global Intention Experiments.

I decided to explore that connection in our global experiments on something besides plants and seeds, and to work with another scientist in order to demonstrate that the results we'd achieved in the leaf and seed experiments weren't an artifact produced by a single lab. I approached a Russian physicist named Konstantin Korotkov, a professor at what is now called ITMO University (the Russian National University of Informational Technology, Mechanics and Optics, formerly St. Petersburg State University). Korotkov had advanced on Popp's ideas and equipment after working out that he could measure this faint light far more easily when he ran an electromagnetic field through it, which excited it hundreds of thousands of times and made it far easier to measure.

At the age of twenty-four, Korotkov, already making his name as a well-established quantum physicist, had become intrigued by the work of Semyon Davidovich Kirlian, a Russian engineer who discovered that when anything that conducts energy, including human tissue, is placed on a plate made of an insulating material, such as glass, and exposed to high-voltage, high-frequency electricity, the resulting low current creates a halo of colored light around the object that can be captured on film. Kirlian made big claims for this light, maintaining that his photographs revealed nothing less than the energy field of a living thing, and the state of this field, or aura, as he came to refer to it, mirrored the state of its health.

Eventually Korotkov came up with a means of improving on this rudimentary system and capturing this mysterious light in real time by stirring up the photons of a living system, stimulating them into an excited state so that they would shine millions of times more intensely than usual. He developed a Gas Discharge Visualization (GDV) device, which made use of state-of-the-art optics, digitized television matrices, and a powerful computer, a blend of photography, measurements of light intensity, and computerized pattern recognition. A computer program would then extrapolate from this a real-time image of the "biofield" surrounding the organism and deduce from it the state of the organism's health.

When we first made contact, Korotkov was fifty-five and a well-known public figure who had lent an air of legitimacy to Kirlian photography and the concept of human energy fields. He'd written five books on the subject, attracting the attention of the Russian Ministry of Health, which recognized the importance of his invention in assessing health and diagnosing illness. By 2007, the GDV device was widely used as a general diagnosis tool and as a means to evaluate a patient's progress after surgery, and the Russian Ministry of Sport had begun to take notice of Korotkov and his machinery, even using it to assess the state of athletes training for the Olympics. Outside Russia, thousands of medical practitioners were using his machines, a fact not overlooked by America's National Institutes of Health; indeed, a portion of Gary Schwartz's grant was to be used to investigate the "biofield" using Korotkov's equipment.

Korotkov is an interesting paradox: a lithe, compact figure with a shaved head who is taciturn and methodical about his work but effusive in his private life. Although humble about his celebrated inventions, he is drawn to the grand gesture, once arriving at a formal event in Japan dressed in traditional Japanese kimono, brandishing a samurai sword. While Korotkov enjoys the notoriety he has achieved with these practical applications, his own private passion is the effect of human consciousness on the physical world, and he is infused with a strong sense of spirituality, which developed after a number of astonishing findings discovered in the course of his work. Although raised an atheist in keeping with the culture of Cold War Soviet Russia in the 1950s and 1960s, he was drawn increasingly to the larger questions of consciousness, particularly the question of how long this mysterious light lingers with the body after death.

In a series of experiments carried out in the late nineties, Korotkov and a team of assistants had taken readings of dozens of newly dead men and women, and found that for many hours, there was no principal difference between the gas discharge glow of live people and the cadavers. Furthermore, the pattern of the light over time followed distinctly different patterns, which seemed to mirror the nature of their deaths; when people died gently, so did their light, but when they died more violently, their light had more abrupt transitions. Those who died a natural death had larger oscillations during the first fifty-five hours after death, which afterward receded to gentler waves.

Although materialists would argue that the light was simply the residual physiological activity of muscular tissues transforming in the process of decomposition, the forensic medical literature made it clear that any electrophysiological characteristics of a newly dead body abruptly changed in the first few hours and either remained constant or created smooth curves. Korotkov's data did not resemble that at all. The only conclusion was that this light carried on after life had ended, evidence of some sort of transition. Korotkov wrote a book about his discoveries and privately became intensely spiritual, regarding this "energy-informational structure" as analogous to what is often referred to as the soul, connected to but ultimately independent of the human body. While carrying on his work for the various Russian ministries, increasingly he was drawn to examining the nature of consciousness, specifically the effect of our thoughts on others.

When I first approached Korotkov to work with me, we decided that our first experiment would be rudimentary: attempting to affect water with our thoughts in some subtle way. One of the subtlest of changes, he suggested, would be to measure any changes in the configuration of water molecules, which we now know have the peculiar ability to act as a team. Two Italian physicists at the Milan Institute for Nuclear Physics, the late Giuliano Preparata and his colleague, the late Emilio Del Giudice, had demonstrated that water has an amazing property: when closely packed together, molecules of water exhibit a collective behavior, forming what they have termed "coherent domains," like a powerful laser light. These clusters of water molecules tend to become "informed" in the presence of other molecules, polarizing around any charged molecule, storing and carrying its frequency so that it may be read at a distance.

In a sense water is like a tape recorder, imprinting and carrying information whether or not the original molecule is still there. As Russian scientists have observed, water has the capacity to retain a memory of applied electromagnetic fields for hours, even days, and other Italian scientists, from Sapienza University of Rome and the Second University of Naples, and more recently, Luc Montagnier, the Nobel laureate and codiscoverer of HIV, have confirmed Preparata and Del Giudice's findings: certain electronic resonance signals create permanent changes in the various properties of water. The Roman and Neapolitan team also confirmed that water molecules organize themselves to form a pattern on which wave information can be imprinted. Water appears both to send the signal and also to amplify it.

As with plants, animals, and people, liquids like water "glow." The GDV machine is sensitive enough to measure various energy dynamics of water, and can detect any change in the emission of light on the surface of the liquid, which in turn depends upon how the water molecules are clustered together. Numerous experiments conducted by Korotkov's team on a large variety of biological liquids demonstrate that GDV equipment is highly sensitive to changes in the chemical and physical contents of liquids, which don't show up in ordinary chemical analyses. His equipment has been able to distinguish the infinitesimal differences, for instance, between blood samples of healthy people and those patients suffering from illness, between natural and synthetic essential oils with the same chemical composition, and even between ordinary water and that which has had highly diluted homeopathic remedies added to it.

For our first experiment, Konstantin would fill a test tube to the top with distilled water and insert an electrode connected to standard GDV equipment. The plan was to measure and compare the signals being emitted from the water before, during, and after the experiment. We would ask my internet audience from my Intention Experiment website, e-newsletter, and social media pages to send love to a photo of this sample of water, an attempt to prove the claims of the late Japanese naturopath Dr. Masaru Emoto that emotion can change the structure of water.

Dr. Emoto had become well-known for a series of informal experiments, published in *The Hidden Messages in Water* and other books, suggesting that our thoughts get embedded in water. He'd asked volunteers to send positive or negative thoughts to water, then froze the water and photographed the ice crystals. Those crystals sent positive intention resulted in beautiful symmetrical shapes, claimed Emoto, whereas those samples exposed to negative intention—fear, hate, anger—formed muddy, asymmetrical crystals. As outrageous as his work seemed, it had been successfully replicated twice by the noted parapsychologist Dr. Dean Radin, chief scientist at the Institute of Noetic Sciences at Petaluma, California.

Still a bit stung by some of our initial technological failures with the leaf experiments, I deliberately limited advertising of the upcoming experiments to our own e-community, so as not to overwhelm the Ning system. Even without much promotion, thousands of people signed up for this next experiment from eighty countries around the globe, including a large showing from every continent other than

Antarctica, and from other far-flung places: Indonesia, Zambia, Costa Rica, China. Word about the experiments had got out—even to Emoto himself, who sent me an email wishing us good luck.

In the evening of the appointed day, Konstantin sent us a photo of the experimental test tube, which we posted on our website but was visible only to those who had registered for the experiment, then turned on his GDV instruments and a CD of Rachmaninoff to keep him company, and waited.

Hours later, after the experiment was over, Konstantin reviewed the measurements his equipment had taken and discovered a highly significant change. The light emissions in the water had gained in intensity, and there was also a significant effect in the change in the total area of the light emissions. However, these variations occurred *before* the actual experiment started, came to a halt six minutes before the planned intention time, and only started up again once we'd finished. When we looked at the comparison between the time of our intention and twenty minutes before that, the significance in the data disappeared.

Perhaps our intention had been too passive or diffuse and might work better if we focused on something more specific, as we had done with the Germination Experiments. After all, the idea of an amorphous emotion like love is highly individual, particularly when it is being sent to a beaker of water. And a number of our participants had managed to get into the site early, which could have skewed the results.

We decided to repeat the experiment on January 22, 2008, but with three important differences: we'd use a very specific intention with our experimental sample, asking our audience to make the water "glow and glow"; we'd set up a control, an identical beaker of water with distilled water from the same source, also attached to the GDV machine; and we'd extend the overall time of taking readings.

This time, we recorded a highly significant statistical difference in the spread of light and its intensity during the intention period and the period afterward, compared with the measurements of the control beaker. Most intriguing, the big change occurred just during the ten-minute window of our intention, compared to before or afterward. Although our participants numbered fewer than the first time around, we'd had a much larger effect. Once again, size of group made no difference to the outcome.

You begin by making certain assumptions, you construct a careful hypothesis, you design a way to test it out, and then you sit back to see where you land, only to find that a few of the confident assumptions you have about the universe have been blown to smithereens.

Of the eleven experiments that we'd been able to run successfully, ten had achieved a successful outcome—all but one of them statistically significant—but in the process, they'd overturned every one of our initial assumptions about how group intention might work.

I tried to unpack what we'd learned about what was happening. We'd been able to change water and plants with our thoughts, whether we were together in a room, all in disparate places, or even thousands of miles away from our target. And our thoughts were affecting things, even though we'd never been sending intention to the thing itself, which of course was sitting in a distant laboratory, but only a *symbol* of the thing: its photograph.

Although the only point of contact was a photograph on an internet site, my participants readily established a profound connection with one another and with the target. Thinking in a group seemed to create a nonlocal psychic internet of instant connection, where the distance between the participants no longer mattered, even when we weren't working with real targets and intention—just their photographic representation, in a sense like a voodoo doll.

When we began the global Intention Experiments, Gary and I had operated on the assumption that it was important to have some live connection with our target, which is why we'd first insisted on a webcam display of the actual target in the early studies. But during both the Germination Intention Experiments and Water Experiments we discovered that human consciousness can connect and affect a *virtual target* and that this connection proves just as potent. As psychics and other clairvoyants have maintained for years, the symbolic representation of something, like a map coordinate, easily enables consciousness to zero in on the target.

The size of the group hadn't mattered either; a tiny group of a hundred in a room in Rhinebeck a thousand miles away from the target had proved as potent as groups five times that size. The Second Korotkov Water Experiment had fewer attendees participating but a greater effect. Distance from the target didn't have any bearing on the outcome either. My Australian audience eight thousand miles away from the target in Tucson, Arizona, achieved the same size effect as did a group of intenders located in the neighboring state of California. When sending thoughts to something, bigger or closer wasn't necessarily better.

Another strange effect was that intention seemed to affect everything in its path; when the seeds were part of an Intention Study, every seed was in some way affected, whether or not it was actually targeted. This also had a big implication, suggesting that living things register information from the entire environment, and not simply between two communicating entities.

What seemed to count most was experience. Our most impressive results came from those who were practiced in sending focused thought, such as experienced meditators or healers. Our most successful Germination Experiments, where the seedlings sent intention in this case grew about twice as high as controls, had involved my audience in Hilton Head, South Carolina, which included five hundred longtime practitioners of Healing Touch. And from both the Germination Experiments and the Water Experiments, we'd also learned that our intention worked better the more specific we'd been.

These first experiments had been rudimentary, even a bit crude, but their implications were enormous. They even challenged certain Newtonian laws that form the backbone of classical physics. Newton described a well-behaved universe of separate objects acting according to fixed laws in time and space, and one of the most fundamental of these is his first law—the idea that any given object remains at rest or continues to move at a constant velocity unless acted upon by an external force. In that law is embodied the cornerstone of many confident assumptions we make about how the world works, the notion that things are static, separate, and inviolate unless something physical, some force—a push, a punch, a swift kick—is done to them. Indeed, all of Newton's laws describe things that exist independently of each other and require some physical, measurable energy to change, even to move.

Very little about our experiments reflected anything we could regard as a Newtonian view of the world. We weren't *doing* anything to an object; we were *thinking* to that object. The effects we were recording were more akin to the rogue behavior of quantum physics, as first defined by Niels Bohr, and his protégé, the German physicist Werner Heisenberg. They recognized a few fundamental aspects of the quantum universe. In the world of the tiny, things aren't actually things yet, but only a tiny cloud of probability, a potential of any one of its future selves—or what is known by physicists as a "superposition," or sum, of all probabilities.

It is now accepted in the scientific establishment that in the hermetic world of the quantum, physical matter isn't solid and stable—indeed, isn't an *anything* yet, and what dissolves this little cloud of probability into something concrete and measurable is the involvement of an observer. Once these scientists actually observe or measure a subatomic particle, this little cloud of pure potential "collapses" into one particular identifiable state.

The implications of these early experimental findings in quantum physics, what is now termed the Observer Effect, have always been profound: living consciousness somehow is the influence that converts the potential of something into something real. The moment we look at an electron or take a measurement, *we help to determine its final state*. This has always held a number of uncomfortable implications, the greatest of which is that the most essential ingredient in creating our universe is the consciousness that observes it—in fact, that nothing in the universe exists as an actual "thing" independently of our perception of it.

Scientists have always shied away from that uncomfortable notion and embraced a more palatable, if improbable, world view: that there is one set of laws for the large and visible, and another for the microscopic, and once those anarchic subatomic things somehow begin to recognize they're part of something big and visible, they start behaving themselves again, according to dependable, logical Newtonian laws.

A few of the assumptions of that confident world view—time and space as inviolate, Newton's first law, even that idea of there being separate sets of rules for the big, visible world and the unseen particle—had had one tiny hole blown in them by these early experiments.

Both the Power of Eight circles and the global experiments were revealing something more—something central about human consciousness and its ability to trespass over the boundaries of objects and other people, even the borders of space and time. We'd repeatedly demonstrated that the human mind has the ability to operate nonlocally, move through walls and over seas and continents, and change matter thousands of miles away. Scientists struggle with the idea, first proposed by German philosopher Immanuel Kant, that the world is not possible without us, but perhaps what is actually meant by the Observer Effect is that we create as we attend to a particular object, focusing on the object in unison and articulating a very specific request together.

Our experiences did not bear out the theories of the TM group that when you are trying for specific outcome through the power of thought, you need a critical mass. A focused and coherent group of one hundred in a single room had the same effect as thousands clustered around the world and connected via the web. We got the same result on the target whether we had an audience occupying the same space or scattered all over the globe but simply joined by the same thought and the same internet page.

In fact, as I was beginning to realize, it also worked with a group of just eight. The intentions worked, I can only guess, because we were all, at that moment, occupying the same psychic space.

The only thing that mattered, the only thing that appeared to be needed, was any sort of group.

Chapter 5

The Power of Twelve

The effects that I was witnessing from the Intentions of the Week could not have been placebo effects. There were babies, fetuses, even, who were getting healed. There were people who were unconscious or weren't told they were the object of intention. Baby Isabella was born in Spokane, Washington, at twenty-four weeks' gestation weighing just one pound four ounces with intestines detached, a strep infection in her stomach, and weak lungs. Two days after the doctors operated on her intestines in an attempt to connect them, she developed an infection, and they had to operate on her lungs for a second time. She was placed on a number of different antibiotics, and a specialist was brought in, who then determined her infection was antibiotic resistant. The doctors attached a colostomy bag. Her case looked almost hopeless.

Her mother reached out to us to nominate her baby for an Intention of the Week. Seven days after we sent our intention, Isabella had another operation and came through amazingly well. Although the doctors had been concerned that the strep infection had returned and they might have to operate again, to their astonishment, her blood levels, the cause of the alarm, quickly reverted to normal. She began developing normally and after eight months was discharged, a totally healthy child. Her mother called it a "miracle."

In May 2009 Jeuline, from Gothenburg, Sweden, was due to give birth, but the baby boy she was carrying was diagnosed with a rare and severe heart defect that was certain to affect the function of his heart and lungs. The doctors were fearful that the baby would not be able to breathe on his own when born because his lungs were likely to be damaged. And even if he was able to breathe, he'd need to be strong enough to go through at least three different operations on the blood vessels of his heart.

Before her due date, Jeuline asked to be put into the Intention of the Week circle. After our group intention her son was born in far better shape than doctors had predicted. They were amazed that he was able to breathe unassisted and his oxygen saturation rose after breastfeeding, since the reverse usually happens in children with heart problems. He continued to gain weight and remained healthy for his operation two and a half months later and afterward continued to thrive.

"Doctors are surprised at how healthy he is and looks," his mother wrote us at the time. "He has been healthier than other heart kids in a similar situation. A very content, calm, and happy little guy."

There was a teenage runaway who was reunited with her parents. Juracy from Mexico wrote us about her sixteen-year-old daughter, who had run away from home. She was failing math, spending all her free time at parties that didn't break up until the early hours of the morning, and socializing with friends her mother disapproved of. Our community sent an intention for the daughter and her mother to be more loving, to communicate more honestly, and to respect each other's differences. After several weeks I received an effusive note from Juracy reporting that her daughter had come home three weeks after the intentions started and they'd begun having honest, heartfelt talks. The daughter also changed her social media pages, which had been very dark and defiant, to powder pink.

I didn't know what we were witnessing here—a healing success or pure coincidence. The fact that the process was working on a baby—and even a fetus—and also people who were unconscious or ignorant that any effort was being conducted on their behalf tended to rule out an expectation effect. Did this have anything to do with some sort of amplified power of group intention?

I'm just reporting on this.
I don't pretend to understand what "this" is.
I'm learning with you.

For years, these were my standard disclaimers during the Power of Eight workshops, my get-out-of-jail-for-free card. *I am not a healer. I'm just the journalist here.* After witnessing so many miraculous changes in people's lives, for a while I even became indifferent. *Ho-hum. Another miraculous healing. Big deal.*

At the same time I became obsessed with trying to find a precedent for these collective healing effects. *Somebody had to have thought of this before me.* Certainly prayer circles are now an integral part of most modern Christian churches. But my Power of Eight and Intention of the Week groups, in some instances, were producing immediate healings. What was it about a group of people thinking a single thought at the same time that was producing such dramatic effects? This ritual must have been discovered and employed in an earlier civilization.

I began casting around for ancient circles used for healing, and began with the most famous circle of them all: Stonehenge, the giant prehistoric ring of standing stones in Salisbury Plain in England.

Archaeologists are still mystified by the actual purpose of Stonehenge and what would compel a Neolithic civilization to transplant eighty-two Carn Menyn bluestones some 250 km from the Preseli Mountains of southwest Wales to their present location in Salisbury Plain, each of the stones, weighing up to three tons, requiring up to thirty men carrying or pulling them along on leather ropes, transferring them to boats that sailed up the River Avon and eventually the Salisbury Avon for their final journey. Many researchers still take the lead from Stonehenge's first archaeologist William Stukeley, who concluded that the Mesolithic structure was a place of worship; as he wrote in the early 1720s: "When you enter the building and cast your eyes around, upon the yawning ruins, you are struck into an exstatic [*sic*] reverie, which none can describe." Others were convinced that the stone circle acts as enormous calendar, as the positions of the stones allow for the precise identification of the summer and winter solstices, which would have been essential for agricultural planting and harvest at a time when no other means of marking the seasons existed.

But in the month before our first workshop, as I subsequently discovered, Professor Timothy Darvill and Professor Geoff Wainwright, two of Britain's top archaeologists, had broken ranks with colleagues after beginning a three-year excavation project and putting their findings together with the unduly large number of bones that had been excavated previously, showing evidence of some sort of traumatic injury.

"The whole purpose of Stonehenge is that it was a prehistoric Lourdes," Wainwright said. "People came here to be made well."

"Initially it seems to have been a place for the dead with cremations and memorials," added Darvill, "but after about 2300 BC the emphasis changes, and it is a focus for the living, a place where specialist healers and the health-care professionals of their age looked after the bodies and souls of the sick and infirm."

Darvill and Wainwright focused upon the stones themselves and the ancient belief that they were imbued with mystical healing powers, largely from the springs and wells in Wales that had poured over them. But I wondered about the power of their arrangement. Their placement had been no accident. The avenue was aligned with the midsummer sunrise, the blue stones forming a circle within what is now a horseshoe arrangement of two rings of stones.

Archaeologists have discovered patches of soil suggesting there may once have been other stones present. Perhaps the healing was thought to be carried out not just by the stones but also by the placement of healers in a circle, with the circle itself meant to have its own healing power. As there are hundreds of ancient stone and timber circles all around Britain, Darvill doesn't doubt that circles played an important part of healing, but he's seen no evidence to prove that the people standing in the circle were as integral to the healing process as the stones themselves.

Throughout the ages, small circles of people have held a special significance in many cultures and religions, from pagan Wicca to mystical Christianity. The Arthurian legend of the Round Table and the medieval brotherhood of the Rosicrucians were said to have incorporated both Arthurian practices and those of the ancient Essenes, the early mystic sect of ascetics reputed to have educated Jesus.

I contacted Klaas-Jan Bakker, the grand master emeritus of the Rosicrucian order AMORC, who explained that the Rosicrucians believe that they employ healing methods first used by the Essenes and taught to Jesus. The closest parallel to my Power of Eight circles was the Council of Solace, members of which are chosen specifically for healing individuals. Members of the council would usually connect with an ill person to ensure he or she is receptive, then send healing thoughts at specified times of the day by getting into a focused mental state and visualizing the person to be healed. In addition to individual healings, at noon every day, in all Rosicrucian Temples a ritualistic Council of Solace ceremony takes place, in which Rosicrucians are to send a positive healing intention for those in need and for the planet. A number of other practices offered some parallels with the psychic internet I'd discovered in our experiments and circles.

As the Rosicrucians claimed to have inherited their practices from those of mystical Christianity, I started exploring more traditional religious uses of the circle.

Many books of the Bible, such as Acts, Ezra, and Jonah, talk of the power of group prayer to conjure up divine guidance and protection and to prevent disaster, and St. Teresa de Ávila allegedly pioneered the use of small prayer groups in the Catholic church. Muslims make the Hajj pilgrimage to Mecca, where they form concentric circles to pray around the Kaaba, Islam's holy center. In Judaism, all synagogues make use of minyans, a group of at least ten (and men, in the case of orthodox churches), one function of which is to pray together for the healing of a member of the congregation. When congregation members recite the Jewish thanksgiving prayer Birkat HaGomel to give thanks for surviving a traumatic experience or life-threatening illness, a minyan must be present. "Minyan" comes from the Hebrew word *maneh*, which is related to the Aramaic word for "mene," or number, connoting a need for a certain critical mass of people. Clearly, prayer groups had been extensively used in most religions.

When studying uses of group prayer in Christianity, I stumbled across an old sermon by the nineteenth-century British Baptist preacher Charles Spurgeon discussing the meaning of certain passages in Acts, the narrative of how the apostles built the early Christian church. Spurgeon focused on Acts 1:12–14, which relates the story of how the twelve apostles of Christ essentially carried out their first prayer meeting. They'd returned from a journey near the Mount of Olives, close to the old city of Jerusalem, and headed to an upper room (some religious historians believe that room to be the Cenacle on Mount Zion in Jerusalem, the place of the Last Supper), where they all engaged in prayer.

Many biblical scholars conclude that the New Testament was written in Hellenic Greek, and, according to Spurgeon, Saint Luke, a Hellenic physician and reputed author of Acts who may have been present during some of the events, chose to use the Greek word "*homothumadon*" to describe their method of group prayer.

Homothumadon is mentioned twelve times in the Bible, mainly in Acts, always to describe the nature of the apostles' prayer. The Authorized King James version of the Bible translates *homothumadon* with the anemic phrase "with one accord," but Spurgeon maintains that *homothumadon*, an adverb, is in fact a musical term, which means "striking the same notes together." Elsewhere it has been translated to mean "with one mind and with one passion," and Spurgeon takes it to mean that the apostles prayed "unanimously, harmoniously, and continuously."

Even that latter definition does not convey the depth of the original, I discovered when I looked up the definition of *homothumadon*. The Greek word itself is a compound of two words: *homou*, which literally translates as "in unison" or "together at the same place at the same time," and *thumous*, which means "outburst of passion" or even "rush along," and is often meant to convey intensity of some sort: "getting heated up, breathing violently," even wrath. When combined, the two words evoke the musical image of, say, a Beethoven symphony, of notes that race passionately along in different ways but blend in pitch and tone to perfect harmony, building to a climactic finish. The word emphasizes that apostles were to pray as a passionate unity, with a single voice. "Here is an overlooked secret of the early church," Spurgeon notes. "Over and over again Luke stresses that what they did, they did together. All of them. United and unanimous."

According to Spurgeon, Jesus considered prayer a communal act.

He wanted his apostles to pray together, with the same thoughts and words—like an intention stated together—and many other historical biblical scholars have concurred with him. The nineteenth-century American Presbyterian pastor and biblical scholar Albert Barnes said *homothumadon* emphasizes that the apostles were operating "with one mind. The word denotes the entire harmony of their views and feelings. There were no schisms, no divided interests, no discordant purposes."

Praying in this manner may have even brought the apostles closer together, with a sense of indivisibility; in both life and in their prayer, the apostles were "knit by a bond stronger than death," according to the nineteenth-century biblical commentators Robert Jamieson, A. R. Fausset, and David Brown. Jesus may have suggested this, knowing that the apostles were about to face a huge struggle in mounting essentially a religious revolution. Seventeenth-century English nonconformist theologian Matthew Poole believed the use of the word *homothumadon* signals their sense of stalwart unity in the face of difficulty, leading to a "great resolution, notwithstanding all opposition and contradiction they met with," which they no doubt faced when setting up the early church.

Many of the church's scholars are convinced that Jesus specifically used this kind of small-group prayer as a blueprint to assist the apostles in teaching members of the early church in the preferred new way to pray, and as a sign of Christian fellowship. British clergyman, dean of Canterbury, and archdeacon of Westminster Frederic William Farrar suggests that Jesus deliberately taught them to pray in this manner to have them move away from "mere individual supplication:

"The disciples had long before made the request 'Lord, teach us to pray' (Luke 11:1), and during the three years of association with Jesus, the form given them as an example may very well have grown into the proportions suited for general worship."

This would suggest that the plan was for the members of the fledgling church to pray as a group, of one mind and heart. More recently, Peter Pett, a retired Baptist minister and university lecturer, argued that this technique of praying as a passionate unity was meant to be used by the entire church congregation.

"The total unity of the infant church is emphasised. Both men and women disciples share an equality not usually known outside Christian circles. They pray together as one. Most of the actual praying probably mainly took place in the Temple where they gathered daily with other disciples of Jesus" (Luke 24:53).

Presbyterian minister and former US Senate chaplain Lloyd John Ogilvie believes that the new Christian "movement" was meant to make use of a new type of communal prayer. "In their early goal to build the church, they devoted themselves to prayer together. More than physical proximity, this meant a spiritual unity."

Prayer, writes Ogilvie, is meant to be carried out in relationship:

> If we want power from the Holy Spirit as individuals, we need to do a relational inventory: Everyone forgiven? Any restitutions to be done? Any need to communicate healing to anyone? As congregations we cannot be empowered until we are of one mind and heart, until we love each other as Christ has loved us, and until we heal all broken relationships.

Some scholars maintain that the Gospels and Acts of the Bible were originally written in Aramaic, Jesus's native tongue. If so, one word that appears is *kahda*, an adverb that means both "together" and "at the same time."

Small prayer circles had been an essential part of the early formation of the Christian church. In fact, small intention circles may have been employed, if not invented, by Jesus Christ.

Many of the references in the Bible about the apostles being "of one accord" mention an act of group healing. In Luke (9:1), Jesus gave his apostles "power and authority . . . to cure diseases" and sent them on their first missionary journey together from village to village in Galilee "to preach the kingdom of God, and to heal the sick," and Saint Matthew also noted that when sending the apostles forth to "the lost sheep of the House of Israel," they were to "heal the sick." In Acts, a "multitude out of the cities" traveled to Jerusalem "bringing sick folks," and "they were healed every one." In his commentary, the eighteenth-century British Methodist biblical scholar Adam Clarke also noted about *homothumadon*: "When any assembly of God's people meet in the same spirit they may expect every blessing they need."

I thought about the words of Clarke, who once wrote this about *homothumadon*:

> This word is very expressive: it signifies that all their minds, affections, desires, and wishes, were concentrated in one object, every man having the same end in view; and, having but one desire, they had but one prayer to God, and every heart uttered it. There was no person uninterested—none unconcerned—none lukewarm; all were in earnest; and the Spirit of God came down to meet their united faith and prayer.

When their thoughts were focused and concentrated and communal.

It may be that *homothumadon* is the state of mind necessary for a healing intention circle that is carried out in Christian churches as a practice without a full understanding of its special power. All this suggests that Jesus understood the power of group prayer and was passing the idea of it on to his disciples. Or maybe, as I believe, he was just trying to say that God is within every one of us, but that power gets amplified in a group.

I looked up the biblical Greek word *ekklésia*, which appears in the Bible some 115 times but is apparently mistranslated in the

King James version as "church." A closer translation is a "called-out assembly or congregation of persons who meet with a specific purpose—a group with a unified purpose, united into one body." Church in those ancient times did not mean the building itself or even a vast organization, but just a small assembly, like the apostles, who were "called out" to meet and pray as a passionate unity.

Jesus's original idea of a "church" may have been something akin to the Power of Twelve. Start with the twelve, learn how to pray together, then spread the word. In an early chapter of Acts, the twelve apostles, after their own prayers together, then pray with a group of 120, which includes Jesus's mother, Mary, and his brothers, and slowly gathers adherents, teaching them to do the same.

In fact, in that same chapter of Acts (1:15–26), the first activity of the apostles, after the Resurrection, was to choose a replacement for Judas. It is commonly assumed that Jesus chose twelve apostles to represent the twelve tribes of Israel, but there may have been an additional reason for maintaining the group as twelve, even if the newest member had not been a witness to Jesus's teachings.

The number of the twelve apostles may have mattered as much as the praying itself.

This "called-out assembly" exactly fit my definition of a healing circle. In fact, *homothumadon* and *ekklésia* are perfect metaphors, I realized, for a Power of Eight group: a batch of individuals passionately praying together as a single entity, thinking the same healing thought at the same moment. When people are involved in a passionate activity like a healing circle, they transmute from a solitary voice into a thunderous symphony.

Chapter 6

The Peace Intention Experiment

By the summer of 2008 I was thoroughly fed up with seeds, leaves, and baby steps in my global experiments and ready to take a giant leap. If a small group praying together in one unified, passionate, single voice created some sort of virtuous entity, how far could we take this healing force on a major scale? I was inspired by my friend Barbara Fields, director of the Association for Global New Thought, who had set up the Peace Project, encouraging peace groups to form in different cities, and also by plans to mark the International Day of Peace on September 2008. I called Gary Schwartz and argued that it was time for us to test whether this group mind we were measuring in the global experiments had the power to heal some major issue in the real world. Let's do something huge, I said. Let's see if we can lower violence and restore peace in a war zone. After all, the Transcendental Meditation people have conducted more than five hundred studies examining whether groups of meditators can have an effect in lowering conflict, which have produced some intriguing results.

"If you are trying to affect something this big, you can't just send a ten-minute intention once and expect it to work," said Gary. So how to proceed? Like any good scientist, Gary's impulse is always to begin by replicating whatever study design had worked in this area before. You should build on the work of the TM organization, he said. One study of twenty-four cities demonstrated that when 1 percent of the population was regularly meditating, the crime rate dropped by nearly a quarter. When the TM organization rolled out the study to another forty-eight cities, they had similar results. They'd also been able to show that when a critical mass of advanced TM meditators targeted their meditation to Washington, DC, in 1993 during an upsurge in violent crime, the crime rate fell.

The organization had even experimented with attempts to lower conflict in the Middle East in 1983, and discovered that the higher the number of people meditating on the conflict between Arabs and Israelis in Palestine, the lower the number of fatalities and all violence in both Israel and neighboring Lebanon.

Over the years, the organization has been dogged by rumors of data fixing, but their research appeared to be thorough and well controlled, taking into account many factors, from weather and the seasons to law-enforcement efforts. We could learn plenty from them. The studies had all been published in peer-reviewed journals and so had been subjected to independent scientific scrutiny. But of course, most of the work concerned the effect of a mass passive activity like meditation, which simply strives for peace within the individual. I wanted to take this a stage further, to see what would happen if a large group of individuals were all deliberately willing the death and injury rate to go down.

It so happened that Gary was familiar with the protocol of study design used by the TM organization, which he believed would provide our Peace Intention Experiment with a crude initial blueprint. Some of the studies had examined the effect of the square root of 1 percent of the world's population, the TM group's idea of the smallest critical mass necessary to effect change, which amounted to seven thousand meditators located in the same place and carrying out daily meditations for a certain set period of time, just as our earlier experiments had done. It made sense to continue the ten-minute window of intention we had established. "The TM studies ran for a minimum of eight days," he told me. "You should do the same."

Before speaking with Gary, I'd written to a contact at the TM organization, who'd been involved with many of their studies, for some friendly informal advice. "The first challenge with this kind of research is finding data sources," he wrote back in early July. "Good daily data on interpretable measures is hard to come by. Most government statistics are monthly at best, and way out of date," he said. "But there are some people doing content analysis of conflict, and maybe you can get their databases." He provided the names of a few possible contacts.

By then I'd assembled my dream team of informal "wise elders," as I referred to them: Gary Schwartz; Jessica Utts, a professor of statistics at the University of California at Irvine; Dr. Roger Nelson, formerly of Princeton University and now director of the Global Consciousness Project; and Princeton's Robert Jahn and Brenda Dunne, of the PEAR project.

In order to show whether an effect is higher or lower than predicted, statisticians often use a trend-analysis plot, a technique in statistics that attempts to show an underlying pattern in behavior, or a departure from that pattern, over a specific time frame. Jessica Utts, an expert in statistical analyses of consciousness research, planned to model a prediction of the likely average violence levels we could expect in the months after our intention, if the fighting carried on just as it had done in the previous two years. If there was a large difference, we'd have a compelling indication that our intention had had an effect.

We decided that the study would run from Sunday to Sunday, starting with September 14 and culminating on September 21, the International Day of Peace. As this was a pilot experiment, our first thought was to deliberately keep participant numbers low, again to keep the website from crashing, but things were coming together quickly and I was confident I could find a target easily, so I decided to tempt fate and get my own Intention Experiment community excited by announcing the event in July.

In order to carry out this activity as a legitimate experiment and not just an exercise in goodwill, you need something almost impossible to find in a war: a very precise means of record-keeping of casualties. That immediately eliminated areas of Africa and the Middle East—and indeed most areas of conflict on earth. I also needed a fairly obscure target—one that no Westerners were praying for—so that any change would more likely be the results of our intentions and not a multitude of other possibilities. What we were dealing with was so subtle that we had to control for any outlandish situation, including the possibility that we might have intention "contamination" from people who were already praying for our target group before we started, which would make it impossible for us to claim that any changes in the target had been solely due to the mental influence of the experiment's participants. After all, Konstantin's first experiment had shown a certain amount of thought "contamination" when participants managed to log on to the website too early.

Jessica Utts wanted to be able to use more than a few years' worth of weekly violence data, starting from two years before our experiment until a few months afterward, in order to have a long statistical tail with which to compare. This meant we were looking for a war in which someone had been carefully counting the bodies, had been counting them for years, and was willing to reveal the actual numbers to me.

Throughout the summer I called and emailed every peace organization around the world suggested to me. I called the Department of Peace and Conflict Research at Uppsala University in Sweden. I got hold of the United States Institute of Peace in Washington. I called centers for peace and conflict management at three universities. Each department had good ideas but little access to data. Somebody referred me to Joshua Goldstein, who had tracked war fatalities in Israel for a month in 2002, and then to a professor at Harvard named Doug Bond, who had set up a system devoted to gathering casualty statistics about America's two wars in the Middle East, but I was unable to connect with either of them. Jason Campbell at the Brookings Institution, a nonprofit public policy organization based in Washington, DC, was an excellent source of published data on fatalities in Iraq, but its reports supplied only monthly figures, when I needed daily or weekly numbers.

The US government's Worldwide Incidents Tracking System (http://wits.nctc.gov), which was essentially tracking every terrorist-related death in the world, only offered information up until March 2008. We'd have to wait about eight months after our event to find out if we'd had any impact. When I decided to phone the organization to see if I could get any more recent data, the website did not have a telephone or other contact information—and neither did the web or directory information. I called all over the State Department in Washington, DC, and nobody seemed to have heard of the organization. I got passed from department to department until finally I got connected to the US government's National Counterterrorism Center, a Jack Bauer–type unit within the State Department. The person at the other end of the phone sounded startled that I'd got through and refused to identify herself but was nevertheless extremely interested in what I was planning to do with two years' worth of casualty information about the Iraq and Afghanistan Wars and why it was that I was so curious about specific terrorist activity. After a few minutes, she refused to speak further about anything other than to demand more information about me and my Social Security Number.

Iraq Body Count, a website publishing a fairly up-to-date daily body count of deaths due to the Iraq War, was manned by volunteers who had their hands full, they wrote, "simply documenting the daily carnage in Iraq."

I was starting to get frantic. It was late August by that time, our experiment was nineteen days away, and I still didn't have a plausible target, a particularly nerve-racking situation, as I already had six thousand people signed up to participate. At that point I still hadn't publicized the event in any major way, because I still wasn't confident the website was going to work if I overloaded it, and I planned to keep it to under twenty thousand participants, but news of the project had already gone viral on the web. A number of large organizations—Gaiam, H2Om, the Association for Global New Thought, the people behind *What the Bleep Do We Know!?* and its official website, the Oneness organization, the Brahma Kumaris, Intent.com—had announced the experiment to their groups, and hundreds more were signing up to participate every day. How could I have thought that it would be easy to find this perfect, well-recorded war?

One of my correspondents suggested that we limit our search to parts of Sri Lanka, where a bloody civil war had raged for twenty-five years. With so much focus on Middle Eastern and Islamic terrorism, this part of the world map had been largely overlooked by America. It might be the perfect virgin target. I could be confident that it was attracting very few Western prayers.

After writing to four more organizations with record-keeping systems with no response, I was on the verge of calling off the experiment when my contact at Uppsala Conflict Data Program suggested that I try to get hold of the Foundation for Coexistence (FCE) in Colombo, Sri Lanka, which had pioneered an "Early Response Database" record-keeping system and had been tracking the casualties in Sri Lanka for many years. As they continuously monitor both areas for daily rates of killings and violence and feed all firsthand information into a database, they would easily be able to provide us with two years' worth of casualties that had occurred before our experiment and regular updates after our intention week. I followed a thread of leads from Brandeis University in Boston to Manchester University in the United Kingdom, and finally to Madhawa "Mads" Palihapitaya, acting director of development for the Massachusetts Office of Dispute Resolution in Boston, FCE's representative in the United States, who directed me to FCE's former chairman, the noted peace activist Dr. Kumar Rupesinghe.

Rupesinghe is the Gandhi of Sri Lanka, a former publisher who helped to found and now chairs the FCE, a humanitarian organization focusing on peace, human security, and conflict resolution. Under Rupesinghe's direction, the FCE had produced a model for dispute resolution and coexistence between the Liberation Tigers of Tamil Eelam, better known as the Tamil Tigers, or LTTE, the well-armed and trained rebel forces, and the Muslims and Sinhalas, the two majority communities of different faiths. By addressing the grievances of all sides, the FCE had helped to lower violence in Sri Lanka's Eastern Province, and as a consequence, Rupesinghe had been trying to persuade organizations and governments around the world to develop similar programs for early-warning systems, coalition building, and burden sharing in civil wars.

Despite these first inroads, in 2008 there was still no end in sight to either the violence or the war. The Tamil Tigers had arisen in response to discrimination against the Tamils in Sri Lanka by the majority populations, and for twenty-five years they'd waged a campaign to create an independent state in the north and east for the Tamil people. Over that quarter century the Tigers had become a smooth-running military machine, pioneering a number of firsts in terrorist activity: the first organization to invent and make regular use of the suicide belt; the first to forcibly recruit children into terrorist activities; the first to choose women as suicide bombers. By the time of our Peace Intention Experiment, they'd notched up well over three hundred suicide bombings, the greatest number of any terrorist organization, and the most audacious of terrorist assassinations, including two world leaders—Indian prime minister Rajiv Gandhi and Sri Lankan president Ranasinghe Premadasa—and an unsuccessful attempt on a third, then Sri Lankan president Chandrika Kumaratunga, which nevertheless resulted in the loss of her right eye. Ten months before our experiment, in an attempt to kill Sri Lankan minister Douglas Devananda, a woman named Sujatha Vagawanam had detonated a bomb hidden inside her bra, an attempt that proved unsuccessful but that was entirely captured on someone's mobile phone and uploaded to YouTube.

Over the years, ceasefire talks had broken down four times, the last in January 2008; after May, the Sri Lankan government had given up and simply decided to exterminate the organization by whatever means necessary. At the height of their power, the Tamils had controlled three-quarters of Sri Lanka's landmass; by the time of our experiment, the government forces had recaptured the East, but violence still waged there, and the Tamils had choked off the entire North of the country, where they maintained their stronghold, displacing more than two hundred thousand people. Some three hundred forty thousand people had been killed in the course of the long conflict, and a half million were presently living in refugee camps. The December before, both the Human Rights Watch and Amnesty International jointly implored the UN Human Rights council to end the civil rights abuses of both sides.

When I described our project to Rupesinghe, he was delighted to share his data without charge. In fact, as it happened, the FCE had begun a house-to-house No to Violence initiative that same month, which was to culminate in a candlelight ceremony to be held on International Peace Day—the evening after our experiment. "We will call upon the entire country to raise a flag in their house, with the symbol of our campaign, and then light a lamp and pray or meditate for five minutes," he wrote. "In the evening there will be mass vigils in the entire country with candles and lamps." He was calling on Catholic bishops, Christian leaders, Buddhist monks, Hindu swamis, and Muslim imams to follow suit and lead their following in prayers. "Since it will be a Sunday, Christians will be going to church, and we are requesting them to ring a bell," he said. "All religions will be asked to toll their bells accordingly." He asked me to ask our network to follow their example and light a candle that final Sunday evening.

I could not believe the synchronicity of our two campaigns ending on the same day. "This," I wrote back, "feels divinely inspired."

Chapter 7

Thinking Peace

I now needed another bit of divine intervention in the form of a new website.

The greatest remaining challenge was to figure out how exactly this experiment should be run on the internet. Ning had been an excellent, low-cost solution to our smaller experiments and our first attempt to use linked server power, but I was not confident that our present configuration could handle an experiment of this size. As before, we planned to carry out the experiment on a platform other than our main Intention Experiment website, with enough distributed network power to handle the numbers. Months before, we'd been introduced to Jim Walsh, the owner of a company called Intentional Chocolate, who generously offered to pay for bigger server power.

Jim had a webmaster in mind to create the site and run the event, and we sent over all our specs, but he then wrote to say that although we had the server in place, his colleague couldn't host it. We were stuck without a website or webmaster. We just couldn't afford the thousands of dollars for the team who had helped us with the early leaf and seed experiments.

It was September 4 by that time—ten days to go. I was again faced with the prospect of calling off the event when I remembered that at a gathering earlier that summer I'd been introduced to Sameer Mehta and his colleagues, a group of seasoned web designers from Copperstrings, a media website running from India and organized by Tani Dhamija, an acquaintance of ours in the United Kingdom. I pulled out their card and got hold of Joy Banerjee and Sameer. When I explained my position, they generously offered to donate the company's time to set up the experiment and run it on Copperstrings' platform, which was large enough to cope with thousands of visitors. I couldn't believe it. This time, the experiment would cost us absolutely nothing.

Sameer and his team created a separate website and sign-up page, but with one major change: his team would create pages that would automatically flip over during the various stages of the experiment to minimize individual computer problems and boost the chances of the highest number participating. And no one would get in prematurely, as had happened with that first water experiment of Konstantin's.

When September 14 finally arrived, a technical person from Copperstrings stood by to help anyone who was having trouble getting through the website's front door. Our *Reiki Chants* track had been timed to play during the ten minutes of our experiment. The majority of people—including me—gained access. I was overjoyed to watch the pages flip over on cue, first to reveal the target, complete with a map showing Sri Lanka, which "suffers one of the deadliest ongoing conflicts on earth," it read.

Five minutes later, the page flipped over again to our intention page, showing a photo of three young boys, a Tamil, a Muslim, and a Sikh, about ten years old, all arm in arm, next to the image of a beautiful waterfall—the perfect symbol of peace restored.

We asked our participants to hold the following intention: "for peace and cooperation to be restored in the Wanni region of Sri Lanka and for all war-related violence to be reduced by at least 10 percent." I'd quantified our request largely because of our experience with the Germination Experiment, which showed that the more specific we'd been, the more successful.

Everything seemed to be working perfectly, but after the first day's experiment, I discovered that a few thousand people had had trouble logging in, because of the sheer size of the audience trying to access the website. More than 15,000 people had signed up, and in the end 11,468 took part. Many thousands more who couldn't log on to the sites joined in after receiving the URL from the Copperstrings tech team. We had participation in more than sixty-five countries, and every continent except Antarctica, with the greatest numbers from the United States, Canada, the United Kingdom, the Netherlands, South Africa, Germany, Australia, Belgium, Spain, and Mexico, but intenders also came from many far-flung quarters—Trinidad, Mongolia, and Nepal, Guadeloupe, Indonesia, Mali, the Dominican Republic, and Ecuador. We had nearly double the square root of 1 percent of the world's population. Because the Apache server again had more requests than capacity, we asked the Media Temple server team to ramp up the capacity for the follow-up sessions, particularly for the final weekend.

The first feedback about the effect of our efforts was alarming. The following week, I read some preliminary news reports and FCE associate Hemantha Bandara's first figures about the casualty rate of the week of our experiment, which suggested that violence had vastly *increased* during our intention week—in fact to the highest it had ever been in the two-year window being studied. Violence levels in the North escalated dramatically right at the start of our eight days of intention. The North experienced a sudden surge of attacks and killings, largely brought on by the Sri Lankan government, which mounted full-on land, sea, and air attacks to drive the Tamil Tigers from their last strongholds in the north of the island. The Sri Lankan navy sank two Tamil Tiger boats during a sea battle that broke out off the northeast coast, and killed twenty-five members of the LTTE's Sea Tigers unit in a three-hour battle off the northwestern coast. In addition, forty-eight rebels were killed in army offensives within twelve miles of the rebel headquarters in the North, and the air force targeted the hideout of senior LTTE leaders. The government also brought the battle to the rebel's stronghold in the Kilinochchi district, killing nineteen rebels and three soldiers. On their side, the Tamil Tiger rebels repulsed an army advance in the northern Wanni region after fighting that lasted four hours, claiming the lives of twenty-five soldiers.

With all this enhanced government activity, the killings and injuries suddenly escalated, with 461 murders and 312 people suffering serious injuries during the eight days of the experiment, compared to 142 killings and 38 injuries the week before.

The government announced that it would refuse to negotiate or offer a cease-fire until the LTTE laid down its arms: it was committed to finally repelling the LTTE in its last stronghold. Aid agencies began to leave the Wanni district because their safety wasn't guaranteed. The renewed bombing raids forced more than 113,000 from their homes. The UN began calling on both sides to stop killing civilians. It all seemed far more than mere coincidence.

Oh my God, I kept thinking. Did we *do this?*

But then in the immediate aftermath of the experiment, both deaths and numbers of people injured fell dramatically, according to the weekly figures being supplied to us by the FCE. The death rate suddenly fell by 74 percent, and injuries by 48 percent. In the short term, when compared with the thirteen days immediately prior to the experiment, the post-intention death rate was not significantly lower. Average killings fell basically to what they were in the two weeks prior to the intention period. However, the injury levels remained down

43 percent from what they were in the months before the experiment started.

That was just the immediate picture. For our data to be meaningful in any way, we needed to take a longer view, backward and forward. Comparing our data with what happened over the two years and then for a month or two afterward would show us if there were any significant shift over the long term, to see if the downward trend continued or if this was as good as it was ever going to get. We also wanted to determine whether intention lasted or if it only affected violence levels for the immediate period after the intention was sent. Was it going to affect the outcome of the war in the Wanni district? The only way to find out was to sit back and wait for a few weeks for events to unfold, while handing over to Jessica Utts the weekly statistics from Hemantha, plus the weekly casualty updates from August 2006 to 2008 for the eastern and northern provinces.

From the FCE's statistics from August 2006 to the start of the experiment, Jessica was able to model a prediction of the likely average violence levels we could expect in the months after our intention, if the fighting had carried on as normal. We then used the data from the weeks after the experiment to compare the model of what *should* happen to what *did* happen over that month. She created a preliminary time series analysis up until the week that ended on September 14, using an Autoregressive Integrated Moving Average (ARIMA) model, which helps to better understand the data and make forecasts for future events, particularly with data, like ours, that don't stay static but fluctuate with a lot of outlier figures.

In late November, Jessica finally produced a quadratic trend analysis, a more complex model that provided a good statistical explanation for the overall pattern of the data up until the time of the experiment and a plausible modeling of what was likely to happen during and after our experiment. It revealed that violence indeed had vastly increased to levels far higher than predicted during the week of our experiment, but then over the weeks after the experiment had plummeted to well below what the model predicted should have occurred during those weeks. Deaths had begun to escalate from the seventieth week of the two-year report, steadily rising nearly week on week to the record high of our experiment, and then plunged a week later back to levels that hadn't been seen since before the fighting had intensified.

Of course, this could all have been coincidence. We had to consider that the increase in violence just happened to occur during the week of our experiment by chance, and that the lowering of violence was simply the calm that often results after a battle. After all, the Sri Lankan government's army had increased in size by about 70 percent that year and also increased its naval forces.

But in the months that followed, the events played out in an even more extraordinary way. From the perspective of these two-plus years, the events during that week in September proved pivotal to the entire twenty-five-year conflict. During that week, the Sri Lankan army had won a number of strategically important battles, which enabled them to turn around the entire course of the war. After our week in September, the army was able to take the fight to the Tigers on their own turf. The fighting turned into face-to-face combat as the army mounted its ruthless offensive in the North.

On January 2, 2009, the army finally expelled the separatist guerrillas from their capital of Kilinochchi. One week later, the army recaptured the strategic Elephant pass and town of Mullaitivu, opening up the entire northern Jaffna Peninsula, where mainland Sri Lanka connects with the northern peninsula, for the first time in nine years, liberating the entire Wanni district, the target of our intention. Those of the Tager Tamil terrorists who remained were wedged into a tiny corner of northeastern Sri Lankan jungle of about 330 square kilometers. After all the decisive wins in September and January, the twenty-five-year, intractable civil war ended in a bloody finish on May 16, 2009, nine months after our experiment.

Did we do this?

Certainly when we started in September, the rebels still had a tight grip on the North, and there was no foreseeable end to war. Although the army had made some inroads in August, even as recently as May commentators believed that peace talks were out of the question. When noting that the highest weekly total for violence and the most decisive battles in the entire twenty-six-month period occurred during our very intention week, Jessica had only two words to say: "Weird, huh?"

I wanted some independent verification that this was something more than coincidence, so I called upon Roger Nelson, the architect of the Global Consciousness Project and a member of our scientific team. A psychologist formerly of Princeton University, Dr. Nelson was fascinated by the idea that there may be a collective consciousness, evidence for which could be captured on REG machines, the modern-day electronic equivalent of a continuous coin-flipper developed by the PEAR team, with a random output that ordinarily produces heads and tails each roughly 50 percent of the time. In 1998, Roger had organized a centralized computer program so that REGs located in fifty places around the globe and running continuously could pour their continuous stream of random bits of data into one vast central hub through the internet. Since 1997, he has been comparing their output with events of great global emotional impact. Standardized methods of statistical analysis reveal any demonstration of "order"—a moment when the machine output displayed less randomness than usual—and whether the time that it had been generated corresponded with that of a major world event.

Over the decades, Roger has compared the activities of his machines with hundreds of top news events: the death of Diana, the princess of Wales; the millennium celebrations; the deaths of

John F. Kennedy, Jr., and his wife, Carolyn; the attempted Clinton impeachment; the Twin Towers tragedy on 9/11; approval polls of Presidents George W. Bush, Barack Obama, and Donald Trump; the invasion of Iraq and the deposing of Saddam Hussein's regime. Strong emotion, positive or negative—even to presidential decisions—seemed to produce a move away from randomness to some sort of order.

I asked Roger to analyze what had happened to the REG machines during our experiment.

Several analyses revealed that the machines were affected within the twenty-minute window of meditations during the eight days of our Peace Intention Experiment, and that these changes were similar to those that occurred during moments of mass meditation in areas attempting to lower violence. But the changes were most striking during the actual ten minutes of our experiment, at the exact time that we were sending intention.

The outcome of the Peace Intention Experiment was compelling but couldn't be considered definitive. As any scientist will tell you, one result like that doesn't really prove much. There were too many variables: the Sri Lankan government's offensive, the natural course of the conflict, the increase and then plummeting of violence. Nevertheless, there was no doubt that the September week the Intention Experiment had been carried out had been possibly the most pivotal of any during the entire twenty-five years of the conflict. The government had gained vital traction in their progress, which enabled them to turn around the entire course of the war.

Did we do this?

Short answer: Who knows?

We would need to repeat the experiment a number of times in order to show that our intentions had affected the war in any definitive way.

But I did make one important discovery. For the first time I had decided to survey all the participants in mid-October to see how they found the experiment, mainly to check how well the Copperstrings website had held up and whether our participants had access to all the web pages.

Another reason for the survey was that I'd been puzzled by the experience of one of the intenders of an earlier experiment. On August 10, 2007, a chiropractor named Tom had written Gary after participating in the August Leaf Intention Experiment at the Los Angeles conference. Tom said he saw the aura of the leaf and saw a change in the glowing at the puncture sites. "I also had a profound ASC [altered state of consciousness]. The entire room got very dark, and the primary thing I observed in the room was the auras of other people. I see auras a fair amount; just the intensity here was significantly different."

At the time I'd dismissed this as wishful thinking (after all, the fellow had said that he saw auras all the time), but his letter had planted one big question that remained in the back of my mind for months. Besides the effect on the target, was any of this also affecting the audience?

When the answers came pouring in from our participants, it was clear that some of the effects were rubbing off on them as well.

Chapter 8

The Holy Instant

It is as if my brain is wired to a bigger network.

One of the experiment's participants had written that on his survey. And thousands more had described a similar phenomenon. These were not the enthusiastic accounts of satisfied participants. These were descriptions of nothing less than a state of mystical rapture. It appeared for all the world that my participants had entered a state of *unio mystica*, the stage of the spiritual path when the self feels a complete merging with the Absolute. It is the moment when, as Saint Teresa de Ávila wrote, we are "cocooned in divine love," when, as an indigenous shaman once put it, "things often seem to blaze," the moment, as Kabbalist mystic Isaac of Acre described, when his "jug of water" became indistinguishable from the "running well." The Sufis and other Islamic mystics, the Kahuna in Hawaii, the Maoris, the Andean Q'ero, the Native American Indians, sages like G. I. Gurdjieff and countless other cultures have all pursued that moment, beyond time and space, where all sense of individuality disappears and you exist in a state of ecstatic union. *The Course in Miracles* refers to it as "the holy instant." It is, in essence, a spiritual orgasm, and a goodly number of my participants, sitting in front of their computers on their own, apparently had just experienced it.

"I felt like I stepped into a palpable stream of energy along my arms and hands which felt like it had direction and force and mass."

"My whole body was tingling and I was having goose bumps."

"I felt a strong current in my body."

"It felt as if everyone was connected to my skin."

"It was like a solid magnetic force field all around me."

"I did not want to leave the experience . . . it felt deep."

"It stopped soon after the experiment."

What on earth was going on here? Either I had momentarily hypnotized fifteen thousand people or the act of a group experience had plunged them into an altered state of consciousness. And the strangest part of all was that my participants had entered into this space effortlessly, simply by holding on to the power of a collective thought.

Most accounts of the *unio mystica* describe experiences that occurred individually, rather than by groups, outside of indigenous ceremony or a charismatic church service, but they aren't as rare as supposed. At the end of his life, psychologist Abraham Maslow turned his attention to these "peak experiences," as he called them, considering them a common element of the human condition and not simply the preserve of the mystic. He vehemently disagreed with historical accounts of these experiences as being otherworldly. "It is very likely, indeed almost certain," he wrote, "that these older reports, phrased in terms of supernatural revelation, were, in fact, perfectly natural."

Parapsychologist Dr. Charles Tart referred to this state as "cosmic consciousness," a term coined by psychiatrist Richard Maurice Bucke. Tart studied the individual characteristics of this state in many cultures and, like Maslow, discovered certain common threads. The saint, the prophet, the mystic, the channeler, the indigenous native had all described the transcendent moment similarly, with certain definable characteristics.

Most mystical experiences include a profoundly physical component—as Bucke terms it, a "sense of inner light," and in the case of the Peace participants, a palpable feeling of energy. Before the experiment began, I too felt an almost unbearably strong energy emanating from my computer, like a powerful force field, but I had dismissed it as my own projection until I read the survey. Many reported physical sensations that were overwhelming: hands tingled, heads ached, limbs felt heavy or painful, emotions felt raw, a powerful, infectious energy seemed to emanate from the computer. Lori, from Washougal, Washington, felt physical sensations in her chest—"an opening." Teresa, from Albuquerque, described the feeling as being part of a power surge, "sort of like what I imagine it would be like to be locked in a tractor beam like is described on *Star Trek*," she wrote. "I was being pulled along on this giant wave of energy while also being part of the cause of the wave."

Participants also reported strange, highly detailed visualizations, almost like hallucinations, and even other sensations, like smells:

"A bright whiteness that shocked me to awareness." (Susan, Wolfe Island, Ontario, Canada)

"A vision of the 'Net of Indira' net of light surrounding the globe, with a beam coming from it focused down to Sri Lanka." (Elizabeth, Port Townsend, Washington)

"Soldiers on both sides putting down their weapons in a pile, then seeing them cultivate their crops peacefully." (Marianne, Bournemouth, United Kingdom)

"A large group of refugees meditating and communicating with the soldiers." (Coril, Pomona, California)

"A clear picture of arrows going back and forth in the dark, then a huge downpouring of love centered on Sri Lanka." (Kathleen, Sonoita, Arizona)

"A slight smell of acai, honeysuckle or vanilla. We have no aromatic flowers in ours or the surrounding yards." (Lisa, Las Vegas)

The majority of my participants had been weeping during the whole of the experiments, not, as I initially assumed, because of compassion for or identification with the Sri Lankans, but because of the power of the connection. "The first day I started sobbing," wrote Diana from New Orleans, "not from sadness but how overwhelming it felt to be connected to so many people. It was POWERFUL."

The fierce emotion, said Verna from Llanon in Wales, came from "the power of the experiment, during Powering Up. I never had experienced anything like it."

Most participants had the sense of not being in control of this experience or even of their own bodies. The energy, the intention itself, and the group situation had come to inhabit them and had essentially taken them over. They were no longer breathing by themselves. Pictures appeared in their minds' eye, they said, that weren't anything they'd made up by themselves. They'd entered an "intense altered state," "set up and ready to be accessed," "a channel for a higher, spiritual power," according to Shyama of New York City. In fact, there was even the sense that you couldn't come back, even if you'd wanted to. "You had to just go with it," said Lisa from Frisco, Texas.

"It sort of came over me. It filled me up and it sought a way out," wrote Geertje, from Lierop, Netherlands.

"It was like I was on autopilot," wrote Lars, from Braedstrup, Denmark. "I performed the experiment and it 'performed' me."

While staring out of the window of the Apollo 14 on the way home to Earth, the late astronaut Edgar Mitchell, the sixth person to land on the moon, experienced a *unio mystica*. It had begun with an overwhelming sense of connectedness, as if all the planets and all the people of all time were attached by some invisible web. He'd had the sensation of being part of an enormous force field, connecting all people, their intentions and thoughts, and every animate and inanimate form of matter: anything he did or thought would influence the rest of the cosmos, and every occurrence in the cosmos would have a similar effect on him. It was a visceral feeling, as if he were physically extending out to the farther reaches of the cosmos.

According to Maslow, when you enter the peak experience fully, with every pore of your being, you leave behind the corporeal essence of yourself. Edgar Mitchell had moved to a place beyond a sense of here and now, and so had the Peace participants: "As always," wrote one veteran of our experiments after its conclusion, "time appeared to stop."

Thousands of Peace Intention Experiment participants described a similar palpable sense of oneness, with all things appearing as a "seamless whole," as William James once wrote. They'd experienced an overwhelming sense of unity with each other and the Sri Lankans—"so intense I could almost 'see' them, most certainly sense them," wrote Marianne, from Bournemouth, United Kingdom, and a welling up of compassionate love, "a flow of energy from the earth and far, far beyond—universal," wrote Gerda from Antwerp in Belgium, even a sense of being pulled "into a wave of light," wrote Ramiro from Texas. It was a feeling of "being light joining with thousands of lightbeams and becoming a vast glowing entity," wrote Filippa, from Mariedfred, Sweden, of "being part of a group mind," said Eoin, from Dublin. The majority reported being overwhelmed by a surge of compassionate love, an overwhelming sense of unity with the other participants, or a powerful feeling of connection with the Sri Lankans.

Maslow also details another phenomenon, a sense of inner knowing, "a direct insight into the nature of reality that is self-validating," as William James referred to it, as though the recipient has gained access to some extraordinary secret key to the universe, a glimpse of which leaves him with a sense of its perfection and a permanent certainty about the future. Bucke described his own mystical experience as a feeling "that the universe is so built and ordered that . . . all things work together for the good of each and all, that the foundation principle of the world is what we call love and that the happiness of everyone is in the long run absolutely certain." There is often a sense of God, but more as the "Absolute" than the anthropomorphic god of some organized religions, and a subjective feeling of immortality or eternity.

In *The Varieties of Religious Experience*, William James described the experience of a clergyman whose mystical episode felt like a face-to-face confrontation with God:

> . . . my soul opened out, as it were, into the Infinite, and there was a rushing together of two worlds, the inner and the outer. . . . The ordinary sense of things around me faded. For the moment nothing but an ineffable joy and exaltation remained. It is impossible fully to describe the experience. It was like the effect of some great orchestra when all the separate notes have melted into one swelling harmony that leaves the listener conscious of nothing save that his soul is being wafted upwards, and almost bursting with its own emotion.

Edgar Mitchell experienced this moment as a blinding epiphany of meaning, a sense that there were no accidents and no opportunity to derail this perfection; the natural intelligence of the universe that had gone on for billions of years and forged the very molecules of his being was also responsible for his present journey. Everything was perfect, and he had his place within that perfection. Many participants in the Peace Intention Experiment felt a similar sense of life's perfection and a bond to all that is. Clare from Salt Point, New York, wrote that it felt like a sense of "Connection!!! To the Universe. No struggle. No doubt. Complete in Stillness." It was, wrote Geertje from Lierop, Netherlands, a sense of certainty, of feeling "connected and at home."

Ultimately, the experience had been ineffable, as though the person had reached a different dimension in the universe that cannot be compared to anything more earthbound. So different was it from any other state of consciousness they had experienced that they did not have the language to describe it, even in metaphor. Ana from Cheriton, Virginia, felt a powerful uplifting of energy, which appeared suddenly in the evening without her doing anything to make it happen. The room then felt "charged" with this uplifting energy. She wondered later whether she'd experienced it just because she'd decided to participate and the "energy" was provided. "It wasn't really analyzable." Helmie from Lierop, Noord Brabant, Netherlands, felt she was "growing and growing so big as I couldn't describe."

Stephen, from Northampton in the United Kingdom, felt not just an overwhelming sense of unity with the other participants, but also a very strong sense of connection to the specific objective of the experiment "MUCH more than 'it will be good if I do this'—almost as though I was, physically, not just engaging in the process, but as though I owned it, was part of it, and it was part of me—quite profound, and hard to describe—far beyond just being 'fully involved.' "

In the book *Ecstasy: A Way of Knowing*, Catholic priest and sociologist Andrew Greeley quotes psychologist Arnold Ludwig's defining characteristics of an altered state of consciousness, which Greeley argues apply to mystical ecstasy, including alterations in thinking; disturbed time sense; loss of control; changes in emotional expression; a change in body image; perceptual distortions, including visualizations and hallucinations; changes in meaning or significance, particularly as regards the mystical state itself, like a eureka moment; a sense of the ineffable; and feelings of rejuvenation. Most of the Peace respondents had experienced most, if not every single one, of these states. It was Greeley's view that anyone undergoing this state is actually afforded insight into a greater reality.

The effect on Peace Experiment participants was something more than the power of suggestion. This was like entering a different dimension.

Chapter 9

Mystical Brains

During the workshops that I began holding regularly, the Power of Eight groups were experiencing a transcendent state identical to that of the large Peace Intention Experiment participants when sending an intention: the same extraordinary physical connection, feeling the essence of the person we were sending healing energy to, the same physical effects on the receivers ("felt tingling in my hands and feet and warmth throughout my body"), the same overwhelming emotions with the senders, experiencing "the palpable, strong sense of beautiful pure giving energy coming from the entire group," the same sense of being "bigger than my body," the same longer-term effects—"physical and emotional sensations stayed with me for hours afterwards"—the same powerful feeling of "coming home."

They talked about unbearable heat and feelings of energy, of being in a deeper meditative state than they'd been in before, of connections with the other members of the group that were more profound than they'd ever felt.

And they were beginning to act "of one mind." During the healing intentions, they imagined the recipient healthy and well in every way, and many would record having the same visualization as other group members, or at least something strikingly similar. In one workshop in the Netherlands, a group was sending intention to heal the back problems of a woman named Jan. Most of the members of the group had an identical and very detailed visualization of the recipient's spine being lifted from her body and infused with light.

Recently, in a workshop in Kuwait, while sending intention to a group member with asthma and hay fever, three of the group members had an identical image of the receiver walking freely in a park without being affected by pollen. And in Brazil, Fernanda had participated in a group sending intention to someone with pain in her left hip. During the intention Fernanda felt intense itching in the same place on her own left hip, and woke up in the middle of the night with pain in the same place. The following day, the pain had vanished. Later that morning she discovered that the receiver of her intention had woken up at the exact same time that night, and by that next day his pain was also gone.

So perhaps the strange physical and mental effects experienced by members of the global Intention Experiments and the Power of Eight groups were being caused by a mystical state. For a time, I thought that my participants were simply describing a coherent brain state achieved by the act of deep group meditation, but I soon abandoned that idea. In the case of the Peace Experiment, no one was connected together. Each of the many thousands had been sitting in front of their computer screens, most of them alone, connected to one another only by an internet site.

I had the who, what, when, where—the essential, easy-to-come-by components of the reporter's checklist—but not the why or how the participants experienced such a profound state of consciousness. I was consumed by the need to find some sort of scientific explanation. Studies that have been performed during mystical states suggest that the brain indeed goes through an extraordinary transformation.

The late Eugene d'Aquili of the University of Pennsylvania and his colleague Andrew Newberg, a fellow of the university hospital's nuclear medicine program, have spent their careers examining the neurobiology of the Holy Instant. As Newberg writes, "We know that gentle contemplative practices like mindfulness meditation predict an improvement in one's mood, empathy, and self-awareness. But Enlightenment is something else, marked by a sudden and intense shift in consciousness." D'Aquili and Newberg carried out a two-year study examining the brain waves of Tibetan monks and Franciscan nuns at prayer using SPECT, or single-photon emission computed tomography, a high-tech brain imaging tool that traces blood-flow patterns in the brain. Newberg discovered that feelings of calm, unity, and transcendence, such as during these peak experiences, show up as a sudden and dramatic decrease in activity in the brain's frontal lobes (behind the forehead) and in the parietal lobes, at the back of the top of the head.

The purpose of the parietal lobe is to orient us in physical space, letting us know which end is up or how narrow a passageway is, so that we can navigate through it. This part of the brain also performs a critical function, possibly the most critical function of all: it figures out where you end and the rest of the universe begins, and it does this by getting constant neural input from every sense of the body in order to distinguish "not-self" from "self." In each study of peak experience, Newberg and d'Aquili discovered that the "you/not-you" dial was turned sharply downward. "At the moment they experienced a sense of oneness or loss of self," writes Newberg, "we observed a sudden drop of activity in the parietal lobe." According to their brains, their Buddhist monks and Franciscan nuns were having trouble locating the borderline between themselves and the rest of the world. "The person," wrote Newberg later, "literally feels as if her own self is dissolving."

Ultimately, the meditators and praying nuns experienced a "total shutdown" of neural input on both right and left parietal lobes, leading to a subjective sense of absolute spacelessness, a "sense of infinite space and eternity" and also a limitless sense of self. "In fact," writes Newberg, "there is no longer any sense of self at all."

With sudden reduced activity in the frontal lobes, logic and reason would also shut down, notes Newberg. "Normally, there's a constant dialogue going on between your frontal and parietal lobes," he writes, but if activity changes radically in either area, "everyday consciousness is radically changed."

In active meditation, where the object is to focus intensely upon some thoughts or particular subject of intention, Newberg discovered that the boundary of me/not-me becomes blurred, but the area of attention, in a sense, takes over. Participants in both the large global Intention Experiments and the Power of Eight groups were also being asked to focus intensely on a particular subject, and in a sense the subject may have taken over their minds.

The left parietal lobe shows a restriction of neural input, causing a blurring of the sense of self, whereas the right parietal lobe, which receives instruction to focus more intensely upon the object of attention, is deprived of any neural input other than the object of intention.

It has no choice, Newberg writes, but to create a spatial reality out of the object of contemplation—the intention for peace, in our case—enlarging it "until it becomes perceived by the mind as the whole depth and breadth of reality" and the person feels completely and mystically absorbed into the object of his intention. Many of our participants had indeed recorded this sense of a mystical union with Sri Lanka.

Newberg is quick to note that this brain activity is reflective of a particular state of consciousness—a signature of it, essentially—not its cause. He distances himself from the strict materialists, who claim these states are entirely induced by the brain, and emphasizes that his scientific research "supports the possibility that a mind can exist without ego, that awareness can exist without self" and that his work simply offers "rational support" for these spiritual concepts and for mystical spirituality.

Newberg's work was amplified by the research of Mario Beauregard, a neuroscientist at the Department of Psychology at the University of Arizona, who used functional magnetic resonance imaging (fMRI) to analyze the brain activity in real time of a group of Carmelite nuns during intense spiritual experiences. The results of these experiments clearly showed that different brain regions, relating to emotion, body representation in space, self-consciousness, visual and motor imagery, and even spiritual perception were activated, producing brain states that are utterly unlike those of ordinary waking consciousness. There was strong evidence, Mario told me, for people being literally out of their mind and into an altered state of consciousness during a mystical experience.

Could the altered state have been set off by the music I'd been playing during all the groups and the experiments? Some studies show that a rhythm like the Reiki chant we played can trigger a mystical state by altering the normal activity of the temporal lobes. But in my own experiments, a good percentage of the participants had not been able to gain access to all aspects of the experiment: the music hadn't worked, they missed the initial Powering Up, or they couldn't gain access to certain pages for the Peace Experiment. And it didn't seem to make a difference to their experience. This meant that the essential element, the thing that must have provided lift-off, was participation in a group devoted to the idea of sending out a prayer in unison.

Why would group thought induce such an extreme transformational state?

Group meditation and prayer certainly promoted a sense of unity among practitioners, but not in the intense way experienced by the Peace Experiment participants. I tried to think of other experiences that might induce that kind of extreme altered state, particularly those where the brain waves of participants had been studied.

One similar situation might be a Pentecostal Church experience, where attendees end up being so taken over that they speak in tongues. The Pentecostal movement, begun in the 1900s, and then amplified in the charismatic churches, now constitutes one-quarter of the entire Christian world. Members of the Pentecostal Church believe if they are given the gift of "tongues," they are endowed with the gifts of the Holy Spirit and are able to heal others and prophesy the future. Church members describe the experience as the words being produced *through them* and not really emanating from themselves at all. The state is usually induced by music and singing in a group setting, within a congregation. As it happened, Andrew Newberg has studied the brain states of a small group of Pentecostal Church members before and after they'd got into a state of speaking in tongues to see if their brain patterns mirrored the monks and nuns he'd studied while in a transcendent experience.

As with his earlier studies, Newberg discovered a sudden drop in frontal lobe activity, but no decrease in parietal lobe activity, and indeed his Pentecostal subjects described their experience as like having a conversation with God, where they do not lose a sense of self, but retain a sense of God's otherness.

Newberg also used SPECT and fMRI technology to study the brain waves of mediums and Sufi masters performing a chanting and movement meditation called Dhikr and found an identical brain signature to that of his monks and nuns: a shutting down of frontal and parietal lobe activity, in particular on the right side of the brain. This brain state would make it easier to access creative imagination, says Newberg, and a sense of oneness. And the larger the decrease in frontal and parietal lobe activity, the more likely the participants would experience all the stages of enlightenment. The greatest changes occurred in the right frontal lobes, the area of the brain associated with negative thinking and worry, which might explain why those experiencing a state of enlightenment often describe feelings of bliss.

Besides these feelings of unity, my participants also had a strong sense of having been part of a profound and significant effort. "I felt that I was important doing something like that," wrote Mónica of Mexico City. They felt hopeful, a sense of "human solidarity," an end to their feelings of isolation, part of a "deep sense of Connection. Placement. Purpose," a "major global project," an "obligation" they should take "very seriously," with a "deep sense of longing" after the experiment was over. "I felt a greater sense of purpose than my small life," wrote Barbu from Greenwich, Connecticut. "I felt compelled," wrote Lynne, a doctor in Seattle, "to do this."

In her classic book *Mysticism*, Evelyn Underhill writes that mysticism:

> is not individualistic. It implies, indeed, the abolition of individuality, of the hard separateness, that 'I, Me, Mine' which makes of man as an isolated thing. It is essentially a movement of the heart, seeking to transcend the limitations of the individual standpoint and so surrender itself to ultimate Reality; for no personal gain, to satisfy no transcendental curiosity, to obtain no other-worldly joys, but purely from an instinct of love.

Perhaps the opportunity to join together with strangers in what is essentially modern-day prayer creates a powerful state of completion for individuals, and it is this that Jesus meant by the idea of praying *homothumadon*. We move away from our isolated state of individuality and enter a pure bond with other humankind—a state that is familiar when felt but rarely experienced in modern times. From a neurological point of view, as Newberg describes it, "When your frontal lobe activity drops suddenly and significantly, logic and reason shut down. Everyday consciousness is suspended, allowing other brain centers to experience the world in intuitive and creative ways." "Decreases in the parietal lobe activity," he adds, "can also allow a person to have intensive feelings of unity consciousness."

The results of the Peace Experiment had been provocative, but ultimately meaningless, unless I ran many more experiments, which I planned to do once the dust had settled on this one and I could gather more resources. But as I began to realize, the question of whether it had "worked" or not appeared to be increasingly beside the point. Perhaps the success of the experiment had nothing to do with the actual outcome.

Intending as a group created what could only be described an ecstasy of unity—a palpable sense of oneness. A cosmic power seemed to work through us, providing a feeling repeatedly described as "coming home." The responses from participants suggested that the group-intention experience breaks down separation between individuals, allowing them to experience the "God consciousness" of pure connection. Many found it profoundly transforming, an opening into a reality they never knew existed.

I could accept that people were moved, even changed by the experience, and felt connected to the other people and the target, but then I began reading answers like these:

"I had very specific healing experiences on all days."

"I have been feeling grounded and even tempered lately. More productive and self-determining."

I had never considered that the experience might have residual effects. The survey answers filled out by participants seemed to be the real point here, and it turned many confident New Age assumptions about the power of intention—with the exclusive focus on the object of desire—on their head.

The true point had nothing to do with the outcome of the experiment and everything to do with the act of participation. Perhaps praying together as a group affords a glimpse of the whole of the cosmos, the closest you can get to an experience of the miraculous. And it may be that this state, like a near-death experience, changes you forever.

Chapter 10

Hugging Strangers

Sylvie Frasca, an Italian-born translator who lives in Roanne, France, always had far too much on her plate, but her work schedule the week before the Peace Intention Experiment started had been so demanding that she had been working all day and through the night. On Monday, the second day of the Peace Experiment, still exhausted from her previous week's grueling schedule, she participated in the day's group intention and suddenly felt lighter and physically much better. This feeling increased dramatically as the week went on and she felt a special bond growing between the other participants of the experiment and the Sir Lankans. On Tuesday, her mood became buoyant and she felt her constant worry about work leave her. By Wednesday she realized her priorities had completely changed. That evening she made a pact with herself: never again would she allow herself to be working night and day as she had done the previous week.

Her relationship with her partner was changing too. Although she practiced Reiki healing, her partner, an atheist who had always prided himself on his logical mind, refused to allow her to do Reiki on him for more than a few minutes. On Thursday, for the first time, he gave himself over to an entire hour-long session. They connected in a way they never had before. The following day they had the deepest discussion they'd ever had about Reiki, spirituality, and Sylvie's father's spontaneous healing from chronic sinusitis. For the first time, their discussion wasn't forced and he was sharing the conversation rather than trying to change the subject.

On the final day of the experiment, knowing that she would be traveling back to Italy from France during the time of the experiment, Sylvie planned to send her intention from the car but asked her partner to light a candle for her and to join with the Peace ceremony himself. Shortly after the ten-minute intention, her partner called Sylvie to say that something really unusual had happened. As he was looking at the picture of the waterfall on the Peace Intention Experiment website, he felt pulled into it by a strong, warm comfortable feeling, and he came away from it overwhelmed by positive feelings. Ever since, he badgered her for more information about the experiment, trying to figure out what exactly had happened to him.

It was already clear to me that the group-intention experience caused some sort of major change in individual consciousness. But an even more complex alchemical process appeared to be going on here. Something about praying in a group caused deep, possibly permanent psychological transformation in many participants and improvements in their daily lives. The experience appeared to carry on for a majority of my participants long after the experiment was over, as though they had been touched by something immensely profound. In fact, many participants were shocked by this "borrowed benefit," as one intender put it. "I wasn't expecting anything personally from this and was blissfully surprised," said Joey, from Yachats, Oregon, about his improved relationships. This ecstatic state seemed to be so powerful that it opened the possibility of individual miracles: healed relationships, major life transformations, healed lives.

Sending intention for peace seemed to have instigated some rebound effect, so that a greater sense of peacefulness infiltrated their lives. Nearly half of the thousands completing the survey reported that they felt more peaceful than usual, and this feeling of peace most affected their dealings with other people. More than two-thirds noted some change in their relationships: more than a quarter felt more love for their loved ones, and another quarter said they were getting along better with people they normally dislike or argue with. They were connecting "more, and more profoundly, with people," "working harder to bridge differences," feeling "more open to people," more willing to make new friends and allow themselves to be loved, and "clearer about which relationships to nurture and which to let go." People who annoyed them before the experiment "seem to be appearing less often" in their lives. "Tashie Delek [a Tibetan blessing] is on my lips often," said one. These peaceful changes seemed almost infectious, affecting other members of their family, even those who hadn't taken part.

But something even more fundamental in the ability to connect with people had been transformed in the participants, some opening of the heart that appeared to be indiscriminate and universal. Almost half claimed to feel more love for everyone with whom they came into contact, and nearly a fifth said they were getting along more with strangers. The experience of joining together with thousands of unknown others for a common purpose appeared to give many people permission to open themselves to people they didn't know—and this readiness to connect carried on after the experiment was over. "Recently, I began saying a quick prayer (to myself) for every person I talk to. I pray, 'God bless you, bless you, bless you, and give you a long, healthy, happy life,'" wrote Frances from Bay Ridge, New York. "I have experienced an increase of peace and love with each person I talk to. I realize, only now, that this practice started after the Intention Experiment."

The Peace Intention Experiment had ignited something so powerful within them that they were able to feel more love toward the entire world.

"My love has deepened for all."

"I feel more interested in conversing with strangers. People seem more attracted to talking with me."

"I can see my peaceful heart resonate outward to others as we come in contact."

"A greater connection to my fellow humans everywhere, and more accepting than judgmental."

"I am braver to show strangers the love I feel, through kind glances."

Some found themselves acting more peaceably in the world in every regard and thinking bigger—"pulled out of my own petty concerns," as Sallie Lee put it. One participant decided to spend the 2008 Election Day in prayer: "As a nation, we are so polarized. Don't feed the hatred of Bush." Another was more easily able to cope with the passage then of California's Proposition 8, which banned same-sex marriage: "I am mightily energized, rather than angry."

The majority reported extraordinary changes in themselves, some capacity to allow for differences of opinion and deal with it in a far more measured way:

"I listen more."

"I accept what is and ask for help from something intangible, to improve or redirect circumstances."

"I am more forgiving. I feel compassion towards others."

"Gained a bit more objectivity about a couple of situations in my life."

"More straightforwardness and honesty."

"More able to express myself in a peaceful way."

"More confident, at ease, and less bothered by the outside commercial pressure."

"Less affected by things said or done to 'egg me on.'"

"Try to allow room for disagreement as an option."

"More aware of inter-office politics and how unnecessary and childish it is."

"Less condemning, and approaching people with an open mind."

"Conscious of unnecessary conflict much more quickly, and give up struggling with others. I honor them instead."

A number reported positive results after experimenting with these changes of attitude in real-life situations. "Last week I dealt with a seemingly dead-end position in business in this way and both parties came out freely and with no giving in, just clearing things," wrote Tony from Dallas. "It felt like a miracle."

For many, participation in the experiment enabled them to be more generous toward themselves as well: "more loving"; "less critical"; "more satisfied with life"; "calmer and more grounded and even-tempered"; "more empowered and connected with circumstances"; "a greater clarity of purpose of the heart." Many now felt both "more at ease with the world" and more satisfied with their own lives and life choices.

"More at peace with myself and more satisfied with my life in general."

"More gratitude for the blessed life I have and compassion for others in the world."

"More 'at peace' even though my external realities, especially financial, are the lowest they have been in a long while."

"I stay close to myself and yet feel more connected with the other."

"A certain background inner confidence on a certain level that I am a being separate from any circumstances I am in."

"A hunger to grow as a person."

For many months I continued to ask myself what it was about this experience that had made it so profound for the participants. Was it the thought of taking part in an international intention for peace? Or involvement in a mass event? The fact that Jesus had advocated group prayer should have been enough of an endorsement, but I still needed a twenty-first-century explanation.

Dr. Andrew Newberg once carried out a survey of more than two thousand people who'd undergone an experience of enlightenment, and discovered that people of all religious faiths (and even many atheists) shared five characteristics, regardless of the path they'd taken to that experience: a sense of unity, an extraordinary intensity of experience, a sense of clarity and new understanding, a surrender to not being in control, a sense that "something—one's beliefs, one's life, one's purpose—has suddenly and permanently changed." Many—indeed, most—of the Peace participants had experienced all five effects.

The transformations experienced by the Peace Experiment participants could have been the aftershock of undergoing such an extreme experience. A number of researchers consider mystical ecstasy one of the most powerfully emotional human experiences. As Abraham Maslow wrote, this is "the way the world looks if the mystic experience really takes. . . . If you've gone through this experience, you can be more in the here and now than with all the spiritual exercises that there are." Certainly there is evidence that a transcendent experience is psychologically good for you. Andrew Greeley discovered that people who had undergone a mystical experience had a new sense of their lives and recorded far higher levels of psychological well-being than those who hadn't had the experience. Newberg had also discovered evidence that people who experienced mystical states have far higher levels of psychological health and enjoy improved relationships, improved health, and a deeper meaning and purpose in life. In fact, in one study, terminal cancer patients who had undergone a drug-induced mystical state enjoyed the kinds of improvements in mental health ordinarily seen with psychotherapy. The scientific literature contains many case studies of patients who'd experienced spontaneous healings of a variety of conditions, even alcoholism, after a mystical experience.

But I kept going around in circles. Could it have been the meditative techniques I'd employed? The rituals I'd asked my intention group to perform had some parallels with the Rosicrucians and even some Western mystical traditions like the Anthroposophical Society. All these organizations made use of distant mental intention, with steps that included techniques not dissimilar to my Powering Up procedure—intense initial concentration, visualization, a heart-centered approach, a very specific request to the universe. In those traditions these techniques were claimed to provide a transport to the divine.

But I was left with one basic question. We had a situation of mass prayer, but the effects on the people doing the praying eclipsed the effects on the target. What past tradition found the same rebound effects? That question led me to Jeff Levin, a professor at Baylor University. Levin is a biomedical scientist and epidemiologist who is also a religious scholar, holding both a distinguished chair and directorship of the Program on Religion and Population Health at the Institute for Studies of Religion at Baylor and an adjunct professorship of Psychiatry and Behavioral Sciences at Duke University School of Medicine. Devoutly religious and a member of a conservative Jewish congregation, he was the first scientist to make a systematic study of the literature on the effect of religion on physical and mental health, particularly the effect of Judaism on health. In addition to his vast study of the effects of religious healing, Levin is also one of the few to ask another very basic question: Does the transcendent experience itself have any health effects? Besides religion, Levin is passionate about healing, has studied all the major esoteric healing traditions, and continues to investigate what exactly it is in the transaction between healer and healee that is responsible for pushing the right button. In the ongoing fractious debate between science and religion, Levin has been instrumental in opening up the conversation.

If anyone could explain what was going on with the Peace Experiment and Power of Eight participants, it was going to be Jeff Levin.

Chapter 11

Group Revision

Jeff Levin had never had heard of anything like the rebound effects experienced by our participants. Nor had Larry Dossey, a medical doctor and author of numerous books on prayer and healing. Nor had any other experts I consulted. Most of the research they could offer me about transformation and healing examines the power of ritual to bring it about. Many esoteric traditions speak to the ability of the mystical experience itself, induced through some sort of extreme practice, to eliminate psychological illness. In the Hindu tradition, the point of yoga (union) is *Samadhi*, the mystic union with the all that is, and some of Levin's research has uncovered the healing power of various yogic rituals to "obliterate" psychological stress and help restore a person to the natural rhythms of health.

He largely reiterated what I already knew about prayer circles: they were used by American Native Americans and other indigenous groups, and modern-day charismatic Catholic and Pentecostal congregations, as well as by many Protestant churches. The closest parallel to the effect of my intention circles that Levin could offer was research he'd unearthed showing that certain rituals for healing could produce changes in emotion, self-awareness, and the sense of oneself and one's capacities among those taking part, and also bring about improvements in neurobiological mechanisms, such as neurotransmitters and immune markers, all of which would make for healthier relationships. But every single one of the healing practices he could think of counted on certain extreme rituals to bring about effects. As noted social commentator Barbara Ehrenreich recounts in her book *Dancing in the Streets*, ecstatic practices have been a central part of most cultures since prehistoric times, but invariably involve elaborate pageantry and ritual.

Even modern secular group rituals such as those used at a camp for Wiccan lesbians or bisexuals who'd suffered sexual violence made use of a variety of rituals: healing charms, chanting, and immersion in water. Deborah Glik, a professor of Community Health Sciences at the University of South Carolina who studied the effects of both New Age and more traditional religious healing groups in Baltimore, discovered that these types of rituals often triggered altered states of consciousness, which themselves had healing effects. Ted Kaptchuk, director of the Harvard-wide Program in Placebo Studies and the Therapeutic Encounter, maintains that rituals of any kind, whether relating to Navajo circles or even Western medicine—particularly those infused with elaborate pageantry, costume, sound, movement, touch, and symbols—create powerful healing mechanisms in the bodies of the participants through "layers of sensations and behaviors":

> Healing rituals create a receptive person susceptible to the influences of authoritative culturally sanctioned "powers." The healer provides the sufferer with imaginative, emotional, sensory, moral and aesthetic input derived from the palpable symbols and procedures of the rituation process—in the process fusing the sufferer's idiosyncratic narrative unto a universal cultural mythos. Healing

rituals involve a drama of evocation, enactment, embodiment and evaluation in a charged atmosphere of hope and uncertainty.

The late anthropologist Roy Rappaport elaborates on this idea. These kinds of practices are so powerfully effective, he writes, because they provide "an evocation of space, time and words separate from the ordinary; a pathway of enactment that guides and envelops the patient; a concrete embodiment of potent forces; an opportunity for evaluation of a new status."

So removing people from their day-to-day environment, infusing them with unfamiliar sounds, rhythms, and ceremony, guiding them through a certain powerful all-enveloping sensory experience, having some sort of evidence of supernatural powers at play, and offering a powerful expectation of change all can prompt emotional healing by focusing emotions on what Rappaport once referred to as "the calculated intensification" of the message. In a way, all these researchers were suggesting that this kind of overwhelming sensory experience works through what is essentially the power of suggestion. The transformational experience is triggered when the mind develops a strong expectation of change. Indeed, Kaptchuk claims that placebo effects are "the 'specific' effects of healing rituals."

None of this offered an adequate explanation of the effect of our Peace Experiment. Our participants had not been taken out of the day to day, but remained in familiar environments, using familiar equipment. Compared to the pageantry of a Navajo healing ritual, the sight and sound we'd provided via our website and "Choku Rei" had been fairly austere. The only symbol we'd offered was a picture of a few teenage boys of different religions, arm in arm—hardly "potent forces" that "envelop" and eventually overwhelm those taking part in a massive sensory experience. There was no emotionally charged environment of potent forces because we weren't even experiencing these rituals as a collective in the same room. Furthermore, we had no "message" to be "calculatedly intensified" and no authoritative figure with "culturally sanctioned powers"; in the Peace Experiment, I didn't have a strong presence on the website. The same was true with the Power of Eight groups in my workshops. Even though the participants were present in the same room, they did not take part in much ritual or work with an authoritative presence. The groups usually ran by themselves, after some initial instruction by me, and when organizing people into circles, I usually distanced myself from the effects. I'd never claimed there would be any beneficial effects on my participants. As far as they knew, their participation in either the Peace Intention Experiment or a Power of Eight group was an entirely selfless act.

So if it wasn't the ritual or a shaman in charge or the sight-and-sound overwhelm, what about some sort of group effect? Could our participants have experienced positive effects from the experience of being part of a collective, which New York University social psychologist Jonathan Haidt calls the "hive hypothesis"? Haidt's theory is that people reach the highest level of human flourishing by losing themselves in a larger group. Haidt had built upon the work of the nineteenth-century social scientist Émile Durkheim, one of the first to study the effect of community upon the individual, who referred to the effect of ritual on a group as "a collective effervescence."

"The very act of congregating is an exceptionally powerful stimulant," writes Durkheim. "Once the individuals are gathered together, a sort of electricity is generated from their closeness and quickly launches them to an extraordinary height of exaltation. Every emotion expressed resonates without interference in consciousnesses that are wide open to external impressions, each echoing the others."

Durkheim also argued that once an individual experienced this state, he enjoyed higher levels of happiness afterward.

This kind of "collective effervescence" was more than evident in mass gatherings such as pilgrimages. Every year, some one hundred million participants gather at the banks of the river Magh Mela in northern India to bathe in its reputedly sacred waters during a pilgrimage at Allahabad. Contrary to every expectation, the attendees record higher levels of physical and mental well-being than normal, despite greater risks to health from communicable diseases, poor sanitary facilities, and the cramped and temporary living conditions of a mass gathering. Researchers from the Centre of Behavioural and Cognitive Sciences at the University of Allahabad found that participants were actually in greater health at the end of the pilgrimage than when they'd set off on it, even when the physical conditions were harsh. The scientists concluded that a key ingredient in the association between well-being and religious belief or practice involves "the collective dimension."

The same inoculation from ill health occurred at Woodstock, the legendary 1969 music festival in upstate New York. Despite the extraordinary overcrowded and unsanitary conditions, bad weather, and lack of provisions, no fights or rioting broke out, and the nearly half million attendees experienced a transcendent feeling of connectedness.

Native Maoris report getting a "fire walker's high," an increase in heart rate and a boost in levels of happiness during mass rituals of fire walking. Even group events using repetitive sounds like drumming are healing, causing a lowering of levels of stress hormones and an improvement in immune-system function, such as increases in natural killer cell activity.

Could our Peace Experiment have created a kind of Woodstock effect?

I had to discard that hypothesis too. None of the hive theories could explain what had happened to our participants. They hadn't been together in the same physical space; they'd only experienced Durkheim's group "electricity" virtually. They hadn't been participating in a healing ritual involving an individual, but one focusing on a target of social change. Despite the fact that this was no true community in any genuine sense, this virtual circle had established a deep connection to one another and gone on to heal many of the important relationships in their lives.

Could major and lasting brain changes caused by the mass group-intention experience have been responsible for these transformations? Certainly meditation and shamanic ritual are known to affect the brain. According to Dr. Stanley Krippner, professor of psychology at Saybrook University in California who has made a vast study of indigenous ritual, shamanic rituals create major neural changes, causing the two halves of the brain to synchronize, leading to better integration between what is generally considered the executive part of the brain (the cortex) and the emotional center (the limbic system) and greater synthesis of thought, emotion, and behavior. Essentially, collective rituals help the brain to become more emotionally mature and so help the person act better toward others.

During transcendental states like deep meditation, the brain becomes more coherent, and the executive part of the brain becomes more adept at decision-making. Neuroscientist Dr. Fred Travis, director of the Center for Brain, Consciousness and Cognition at the Maharishi School of Management, discovered this after carrying out electroencephalogram (EEG) readings of the brains of meditators. As with shamanic rituals, after this kind of experience, your brain gets more organized, and various parts of the brain—the intuitive center and the cognitive forebrain—communicate better with each other. Cosmic Consciousness spills over into everyday life, and the brain becomes better at navigating one's world. This could explain why participants recorded such improvements in their relationships. The physical act of intending as a group, which starts off as a mass meditation, induces a state that would essentially train the brain to improve reactions to other people and help individuals to get better at the business of relating. However, these kinds of effects are measurable usually after months or even years of regular meditation. The radical changes in our Peace Intention Experiment participants were showing up after eight days of ten-minute intention sessions.

Acting in synchrony, engaging in a common intention statement or even in the same rhythmic activity with a shared intention, leads to greater social bonding, and this in itself can have a powerful healing effect. Harvey Whitehouse, a statutory chair of social anthropology at the University of Oxford and a leading expert in the science of religion, has written that a ritual involving a "high-arousal cluster" historically has led to the formation of small cohesive communities. Ample evidence shows that being part of any sort of community enhances healing. When I studied stress and its effect on illness, I discovered that the greatest cause of psychological or physical illness is a feeling of isolation—from others, from our family, from our God. Consequently a potent healer of stress in any area of life is simply establishing a strong connection. For instance, in research examining the level of stress among those experiencing financial hardship, even Latino Americans in the lowest income brackets experienced far less depression so long as they regularly attended church and participated in the religious community, and elderly Latinos remained more robust, no matter what the level of poverty, so long as they lived in tight-knit neighborhoods.

I felt I was getting closer to a big first clue about the cause of the amazing transformation experienced by Peace participants, and it had to do with exactly what is meant by "health" and "healing." Many of these esoteric traditions define health as something far greater than an absence of anything going physically awry. For them, true health requires pure and total integration into the whole, a sense of absolute connectedness, while illness results from a sense of alienation from that source. This suggests that genesis of most disease is not the smaller stresses of our lives, but the stress generated by our global response to life: how we perceive our place in the world, particularly within our immediate environment. If that is the case, connecting with others for a singular purpose would be profoundly healing, a palpable reminder that you are part of a greater whole.

Believing in something greater than yourself has a healing power in itself. In his research on many religious faiths, Jeff Levin has discovered that the central tenet of all different traditions is that the universe is not subject to random processes but possesses a divine order, and this belief in a divine plan, a sense that there is a purpose to everything that unfolds in the world, itself is powerfully transformational in every regard.

Those two thoughts were echoed repeatedly by the Peace Experiment participants:

"My mantra for daily life is: I am free, I am Love, and this is what my intention is for everybody I met, because we are ONE GLOBAL FAMILY."

"I sense a unity of being—a common thread which weaves us together in truth."

"I feel encouraged about the future—that this type of activity can create a powerful change for the better."

"Now I believe that there is hope for my country."

"This changes my view on me and life. It gives me hope for a better world."

"I feel more empowered. I feel more connected with the planet."

"More positive about the changes taking place in the world and the direction that we are moving towards is one of unity."

"Hope. Filled."

Maslow distinguishes between two types of mystical experiences—the "green" type—which is a transient peak experience of ecstasy—and the "mature" type, which produces a more long-lasting transformational shift in the person. "In its mature form, [the transcendent experience] represents an ultimate expression of self-actualization and social integration," Jeff Levin writes. "In this context, definitions of health, spiritual well-being, and personal development seem to intersect." From that perspective, Levin believes that the "mature" type of mystical experience, as Maslow defines it—the kind experienced by my participants—is a powerful physical and psychological preventive because it represents the ultimate sense of belonging and purpose in the world. Many experts on the mystical experience concur that the transcendent experience has the power to make permanent changes in the individual in every regard, leading to a sense of a calling to do something altruistic. Maslow says that those undergoing this experience invariably feel an "all-embracing love for everybody and everything, leading to an impulse to do something good for the world," and Evelyn Underhill describes it as being "called to a life more active, because more contemplative, than that of other men." The Indian guru Sri Aurobindo once claimed that "bringing down" the "supermind" and "overmind" not only proves healing to the recipient but also to his entire environment.

After the Peace Intention Experiment, a goodly number of our participants felt compelled to change something major in their lives.

"I applied to the Peace Corps."

"I signed the nonviolence pact."

"I feel more committed to regular monthly mass meditation."

"I am setting up a team to develop a formula for peace based on my understanding of Sri Lanka."

"I immediately started looking for a place to start my own private practice in energy medicine. Quit my hospital job."

It might have been that a large-scale Intention Experiment acts as a potent reminder of what we're supposed to be. The feeling of perfect integration, symbolized by a giant circle of strangers all praying together, offers a sense of purpose—the universe "has meaning and order and that he or she has a place in that order," as social scientists Deirdre Meintel and Géraldine Mossière from the University of Montreal once put it. I wondered if the healing effects my participants had experienced had mostly to do with a feeling of perfect global trust, that so rarely experienced sense that life truly loves us.

Perhaps the key to the healing effects was indeed this sense of trust. James W. Pennebaker, chairman of the Department of Psychology at the University of Texas, has spent more than three decades studying the power of trusting others. His research reveals the power of "opening up." People who can trust others enough to be vulnerable enjoy improvements in immune function, autonomic nervous system activity and psychological well-being, and fewer visits to the doctor.

I witnessed this especially in our Power of Eight circles, where participants revealed intimate details of their health issue to one another. Daniel described what happened with his spine; Rosa, her struggles with her thyroid. In that circumstance, both the discloser and the recipient of someone's secret life require a level of trust that in itself can prove healing. Pennebaker has also studied the social dynamics of opening up and sees it as the primary driver in therapy, perhaps even the primary driver in prayer circles. Disclosure is all about telling someone our story, said Pennebaker.

If so, what we were doing in our healing circles was editing the story, rewriting the details together, offering the possibility of a more positive ending. Perhaps that group process of revision, the idea that you could rewrite a life story or even the story of a wounded country, proves healing to everyone, both the protagonist and those with the pencil in their hands.

Chapter 12

Holy Water

I got bolder in my experiments with the whole psychic internet idea. If we were making all these virtual connections during the large-scale experiments, and it didn't matter whether we were in the same room together or scattered around the globe, I wanted to see how far we could take the virtual effect with a Power of Eight group. I'd seen some evidence of the power of the virtual connections in our Intentions of the Week, but what if we created smaller, virtual Power of Eight groups whose only point of ongoing contact was a telephone line?

I began playing around with groups assembled during teleseminar courses and conference calls. The individual members of these groups were scattered in different geographical locations all over the globe, connected only by a Maestro teleconference master board, which has the capability of dividing the audience into pairs or groups of any size, who then only hear one another and can carry out conversations or exercises by themselves. I would run ninety-minute interactive workshops over six successive Saturdays, offering instruction on the call to the entire audience, after which Joshua, my emcee during these events, would assemble them into small groups of eight to carry out interactive intention exercises, just as the Power of Eight groups did during my workshops.

The fact that these teleseminar groups were not physically present with one another did not make one bit of difference. The virtual Power of Eight participants reported the identical effects as the groups who had been in a single physical place in my workshops: the connection, the physical effects, the overwhelming emotions, the sense of being part of something much greater than themselves, the same intensity of everyday experience felt by those who'd undergone a *unio mystica*. "I felt so sure and confident about everything, that all is possible," said Simone, after being the target of a virtual Power of Eight group for her thyroid. "And everything I was looking at was so beautiful, the trees were such a beautiful green, the asphalt on the road a beautiful grey. And I just felt this certainty." And there were the same improvements with husbands and wives, sisters and brothers, the same desire to connect more with strangers, the same physical improvements felt by both recipients and senders:

"I have intermittent neck pain due to past whiplash injuries. During the session I felt the pain intensify and then the pain was gone."

"I have a tremor in my right hand and the tremor was significantly improved after I received the healing intentions."

"I often have backache or pain in the hip joint. I haven't felt this pain for the past two weeks."

"My anemia has cleared and anxiety has gotten better."

"My left knee does not hurt when I cross it."

"My blood pressure was very high and has steadily gone down."

"The pains in my chest went overnight."

"Less gas, bloating and improved overall symptoms of IBS."

"While sending wellness to the receiver, I noticed energy in my head, jaw and neck, the next day the pain that I have been experiencing in my jaw was greatly minimized and has remained that way."

Virtual Power of Eight group members were also reporting the same kinds of aftershocks that the participants in the global Intention

Experiments had: healed relationships, self-forgiveness, healed lives, a renewed life purpose. In an early meeting with her group, Hilde Palladino from Oslo, Norway, asked for a book deal with a major publisher in Norway to publish her debut young adult fantasy novel. A month later she signed a deal with one of the country's biggest publishers.

When her online group was first assembled, Hilde avoided telling them about a variety of health issues relating to cancer, until another group member revealed information about her own health, and Hilde then felt safe to share. Although she was still undergoing immunotherapy as a mop-up exercise after chemotherapy, at first she only told the group she had allergies because she didn't want her group members to think of her as a cancer patient. Eventually, Hilde shared her history with the group and asked them to include her health concerns in their intentions for her. Within a month, Hilde was able to come off the medication she was supposed to be on for the next ten years. Her doctors told her that she tolerated the cancer treatment remarkably well and was back to full health in record time. "I have more energy, strength, endurance and my health has become very stable," she says.

The group even helped her out of a short-term financial hole. On a business trip to Shanghai, Hilde decided to treat herself and booked a five-star hotel online, even though it was far more than she could afford. Nevertheless, Hilde enjoyed her stay and paid the steep price with her credit card.

After returning home and suffering a few financial setbacks, Hilde examined her accounts and realized she was about $1,500 short. She decided to send an intention with her group to recoup or receive $1,500. Sometime later, Hilde discovered a giant disparity between the current online price of the hotel and the price that she'd originally booked and paid for. She called the hotel to ask for an explanation. The hotel staff member replied, "We obviously made a mistake and will therefore reimburse you the difference of $1,500."

Juliette from Toulouse in France experienced a series of synchronous events after asking her group to set an intention to coordinate her rehabilitative therapy for a spinal injury with the demands of her job and creation of her own new business. Juliette had located a particular chiropractor in Spain known for helping her type of situation, but didn't have the means to regularly travel there and stay in a hotel. As it happened, a friend of Juliette's, with whom she was working on a project, was located in the same area in Spain as the chiropractor and spontaneously invited Juliette to come stay with her for as often as she would need. Then her brother, from whom she was largely estranged and maintained minimal communication, contacted her out of the blue and invited her to stay at his apartment in the Costa Brava. His proposed dates happened to be the very week for which she'd booked some vacation time from her job but had no plans. That invitation enabled her to get to her chiropractor and her project partner more easily, swim regularly (a necessary rehabilitation), and, best of all, reconnect with her brother and his family.

Daphne had been a professional pianist but had to give up playing after being diagnosed with Parkinson's disease six years before. She asked her group to set an intention for her to play the piano and write with ease once more. A few months later, although not back to her professional playing standard, she'd now begun playing "expressively" for an hour each day. "And this year," she said, "I wrote messages on my Christmas cards!"

All of this was fascinating, miraculous even, but much as I was beginning to doubt that the double-blind, placebo-controlled trial could tell me much about what was going on with the Power of Eight circles, virtual or otherwise, even at this point I remained wedded to the scientific method. I still believed that nothing would be considered significant unless I could offer up some more proof in the laboratory. To show we were changing something basic in people, we needed another, simpler kind of global experiment that we could easily quantify. That possibility arose with a strangely formal phone request—"Dr. Masaru Emoto would like to request the honor of a meeting with you in one hour"—a few moments after I'd arrived at the Marriott in Hamburg, where I was due to speak at a conference the following day.

Emoto considered himself a "missionary" of water, believing that water had an intimate relationship with our minds and that by healing water we would heal the world. We had never actually met, although we knew about each other's work. In fact, he had once asked me to stand in for him and speak in his stead at a Spanish conference when he was ill and had introduced me in a video taken from his hospital bed as his "twin soul."

Dr. Emoto arrived at the restaurant bar surrounded by a swarm of aides and a translator. After an elaborate formal introduction, he shared his bold idea with me. He wanted to take his work one stage further by holding a Water and Peace Global Forum on March 22, 2010—World Water Day as designated by the United Nations—at Lake Biwa, the "Mother Lake" of Japan. One of the world's oldest bodies of water, Biwa provides water to 14 million Japanese residents, but since 1983, after rapid urbanization of the surrounding land, domestic and industrial waste has changed the population of microorganisms in the lake, causing outbreaks of red tides, water bloom, and waterweeds. During the conference, he'd hoped to enlist me to hold an international Intention Experiment to show that healing thoughts could help to purify the lake's highly polluted waters.

The ceremony at Lake Biwa was meant to be highly symbolic, said Emoto, a simple demonstration that we could take a step toward the solution of these water-related problems by "considering the aspect of water which is connected with our mind, thoughts, and emotions."

I liked the sound of his idea, but I had some major misgivings. I'd had my own big plans to attempt to purify polluted water; in fact, Gary Schwartz and I had been kicking around the idea of taking a biological organism that appears in polluted water, such as harmful bacteria, and transforming it into its benign form. This was not as far-fetched as it seemed. A number of laboratory experiments had demonstrated that positive intention could encourage the mutation of harmful *Escherichia coli* bacteria, and negative intention could impede it. Even bacteria were highly sensitive to the power of a thought.

But my greatest worry had to do with the challenge we were taking on. The most common substance on the planet, which also contains the world's second most common molecules (after H_2), water continues to bedevil scientists, even those working with it every day in the laboratory. This seemingly simple molecular structure—two atoms of hydrogen for every one atom of oxygen—belies its singularity. Water is a chemical anarchist, behaving like no other liquid in nature, displaying no fewer than seventy-two physical, material, and thermodynamic anomalies, with many more apparently still to be unmasked. Water is among the most mysterious of substances, because it is a compound formed from two gases, yet is liquid at normal temperatures and pressures. It is the lightest of gases and far denser as a liquid than as a solid. Hot water behaves differently than cold water; it freezes faster than cold water does, and ice density increases as you heat it up but shrinks on melting. Water has an unusually high melting point and boiling point. The list of bad behavior goes dizzyingly on.

It is most of what we're made of (humans are about 70 percent water; plants, 90 percent), there are a hundred times more molecules of water inside us than all the other molecules put together, it covers three-quarters of the planet, and life on Earth is impossible without it. But we're still no closer to understanding exactly how it behaves. Attempts to model water continue to fail. You could spend your entire career—and many scientists do—playing around with water and feel like you are getting nowhere.

Much as Lake Biwa was the perfect target for the first live experiment to heal an aspect of the global environment, I wanted to try out a few more basic experiments before launching into such an undertaking for one good reason: Up until that time, the only three Intention Experiments that hadn't delivered a positive outcome had all concerned water.

Our first forays into water experiments with Konstantin Korotkov had led me, through a mutual friend, to Rustum Roy, a materials scientist at Pennsylvania State University, and arguably one of the world's experts on water. Rusty Roy's qualifications were unquestioned; he'd written more than six hundred papers on everything from glass ceramics to diamond films and nanocomposites, and at the time I'd met him, he was the most senior member of the US National Academy of Engineering. *Newsweek* had once referred to him as the "leading contrarian" among American scientists; after one of his impassioned testimonies, the US House of Representatives' Committee on Science, Technology, and Research had given him its first standing ovation in sixteen years.

A typical enthusiasm was a seminal paper on his theory of structured water. Rusty and his coauthors had synthesized all current research at the time for the structure of water, and concluded that those little H_2O molecules are themselves the chief instigators of water's anarchy, in the way they choose to cluster together.

When applied to water, "structure" refers to the position in three-dimensional space and the molecular arrangements of individual water molecules of H_2O, which cluster together like endlessly varied reassemblies of Lego. These clusters remain stable for anywhere from a part of a second to several weeks. It was Plato's belief that water should be represented as an icosahedron, or twenty-sided shape, and 2,500 years later, a few frontier scientists finally began to agree after discovering that clusters of these molecules aren't uniform in any given sample of water. Hot samples have a different Lego shape than cold samples, for instance; some water contains molecular clusters of up to several hundred molecules apiece. It's been discovered that small clusters can cluster even further, creating up to 280-molecule symmetrical clusters and interlinking with other clusters to form an intricate subatomic mosaic.

As Rusty explained to me, the "glue" making these water molecules momentarily adhere to one another is not simply the bonds between the hydrogen atoms, but has to do with a wide range of very weak bonds that exist between the different Lego shapes. These are known as van der Waals bonds, so named after Dutch physicist Diderik van der Waals, who discovered that forces of attraction and repulsion operate between atoms and molecules because of the way that electrical charge is distributed, a property that allows certain gases to turn into liquids.

"It is this range of very weak bonds that could account for the remarkable ease of changing the structure of water, which in turn could help explain the half dozen well-known anomalies in its properties," Roy wrote me. "In its subtler form, such weak bonds would also allow for the changes of structure caused by electric and magnetic fields and by radiation of all kinds, including possibly so-called subtle energies"—like thoughts.

The idea that water molecules get structured is by no means universally agreed upon, but as Rusty argued convincingly, it is structure, not composition, that largely controls the properties of a substance, and if structure is changed, it can completely change the substance without any change of composition. A perfect example of this is diamond and graphite. Both share identical composition, yet diamond is one of the hardest substances on earth, and graphite one of the softest. Their difference is entirely dependent upon which and how many molecules decide to bond.

When I met him, Rusty had just been filmed for a documentary about water, in which an artist had created a graphic illustrations of what structured water might look like. Ordinary water was depicted as separate asymmetrical clusters of molecules floating alone, like wheels with a few spokes blown off, but in the artist's representation of structured water, the molecules had formed two concentric perfect circles. With structured water, the molecules behaved themselves, like a group of orderly schoolchildren seated at a round table.

According to Rusty, research had shown that structured water could be produced through various forms of energy: heat, light, sound, radiation, and, Rusty believed, the power of a thought. As strange as that sounded, there'd been some precedents. Canadian studies had shown that when water used to irrigate plants is sent intention by healers, the hydrogen bonding between the molecules changes in a similar manner to that which occurs in water exposed to magnets, and Russian research demonstrated when healing is sent to a sample of water, the hydrogen-oxygen bonds in water molecules undergo distortions in their crystalline microstructure.

When we were first kicking around how to design the experiment, Rusty told me that structured water is found in the cytoplasm of healthy tissues in the body and may make them healthy because it has a high solubility for body minerals, but it's is also found in healing waters. "This appears to be the structure shared by very different healing waters from some healing spas to silver Aquasols used worldwide," he wrote. Let's test whether we can change the structure to be more like that of the water in those spas, he said.

It took me a few moments to fully grasp what Rusty was proposing. Our experiment was going to attempt to turn tap water into the equivalent of a fountain at Lourdes.

We planned to run our water experiment with Rusty's laboratory team at Pennsylvania State University. Although Rusty and other materials scientists had been hampered in their attempts to find equipment that could correctly demonstrate changes in the structure of water, he believed that a Raman spectrometer would be sensitive enough to capture it.

In 1928, a physicist named Chandrasekhara Venkata Raman had discovered that when light is transmitted through matter, part of the light scatters randomly, and a small portion of this light has different (usually lower) frequencies than the light source. The Raman effect is usually caused by a subtle change in the vibration of a molecule, and this reradiation process offers important information about how the water is structured—the vibrational states of the hydrogen bonds relative to the oxygen in water, for example—and by examining changes in the intensity and shape of the molecular bond, scientists can uncover any structural alterations. I was particularly keen to use Raman spectroscopy because employing a system of measurement that is universally recognized by the scientific community would make our results unassailable.

Water molecules are always moving, like a person constantly lifting hand weights in a gym in a variety of directions. Imagine the single oxygen atom present in every water molecule as your head and the two hydrogen atoms as your arms. The vibrational directions resemble your lifting the hand weights and moving them rhythmically away from your body in front or to the side, or even in a scissor movement above your head, with your arms moving alternately or at the same time. When immersed in water, the Raman equipment sends out a laser light and then measures the number of photons of infrared light that make it back to the detector and are "counted." Usually the energy of the laser photons is shifted up or down, according to the various ways in which the hydrogen "arms" are moving, which is generally represented by four visible "bumps" on a graph.

In deciding on this equipment, Rusty had been inspired by work at Tsinghua University in Beijing measuring the effect of qi, or life energy, sent by Dr. Yan Xin, China's best known Qigong grand master, to a water sample in the university lab from a location a thousand kilometers away. After Xin sent intention, there'd been inexplicably large energy peaks resulting from long-wave far-infrared light waves, suggesting that Xin's qi had definitely affected the water's molecular structure.

After reading of this experiment, Rusty played around with a few preliminary experiments of his own involving a Qigong master and had managed to produce some inconclusive results with a pH meter, convincing the team that if they were trying to test energies involved with healing intention, they would need more sophisticated equipment.

For our own experiment, Rusty's scientific team prepared four glass beakers, labeled A, B, C, and D, three of which were to act as our controls, then filled them with water. Beaker A was placed in the lab room down the hall, beaker B in a mu-metal box, to shield any magnetic fields from affecting the water sample, and beaker C six feet away from the experimental setup. The team then placed the Raman probe into beaker D of our water sample and took measurements every ten minutes for an hour. A long cable connecting the probe to a highly sensitive CCD camera on the instrument would sense the weak Raman scattering from the molecules as they vibrated in response to a red laser light trained on the water sample. The team took identical measurements of the three control beakers before and after the hour of measuring our target beaker.

The idea of changing the structure of water is so abstract a notion that it is difficult to construct an intention that a lay audience would readily grasp. It can't be reduced to a simple idea like the "grow" instructions we had with our Germination Experiments. The Penn State scientists recommended that we show our audience a graph with a blue curve on it, representing the kind of molecular vibrations present in "normal" water—a flat line with a single small ripple that sharply rises to a shape resembling a camel's double hump. We asked our participants to intend for the measurements of the water to be reduced so that they resembled another green line following the same double hump as the blue line but shifted lower on the graph, supposedly representing the molecular vibrations of water considered "healing." Essentially, we were asking our audience to dim the light reflected back from the water.

When Rusty and his colleagues, Dr. Manju Rao and Dr. Tania Slawecki, first examined the results gathered from the Raman probe, they discovered that the light in our target beaker indeed had dimmed. This showed up as a significant drop in the intensity of the camel's double hump on the graphs, with the first big changes occurring during our Powering Up period and the start of our ten-minute window of intention. This downward shift in the direction of the graphs approached the target green line during the time of the experiment, and the sample then returned to its original state an hour later, entirely correlating with the timing of our intention. These sorts of changes were not recorded in beaker A (down the hallway), and although there had been effects recorded in beaker B (in the mu-metal box) and C (six feet away), their changes were not as large as those of our target beaker.

Nevertheless, the scientists were left with a few questions about the equipment itself that prevented us from calling this an unqualified success. When the Penn State team had carried out several of their own studies with Qigong masters and healers, they'd discovered that the healers were emanating radiation also picked up by the Raman spectrometer. With Rusty's own study involving the Qigong master, the pH of the water had gone into strong oscillations shortly before his arrival, although the scientists were not sure whether that was the effect of the qi or due to some instability in the equipment.

We couldn't rule out the possibility that we were using the wrong equipment, or even that the available equipment wasn't sensitive or stable enough to measure something like structured water.

Environmental factors could also have played a part in our results. At 5 p.m. an intense thunderstorm hit the area, and Tania wondered whether it had something to do with the strong changes recorded. "Perhaps as a result of the falling barometric pressure and excess free ions, our deionized water samples were not as stable as they should have been," she wrote me.

Or it could just be that intention doesn't affect the structure of water, or even that the structure of water doesn't change. The changes we'd recorded, said Tania, were like looking at a beam of light from a flashlight on the wall and watching it grow dim at exactly the time we were sending intention. If the battery in the flashlight begins to run out of charge, the beam of light on the wall will get a little dim. It doesn't mean the wall has changed. We had to allow for the possibility that our original hypothesis was simply wrong. And we had no way of knowing which of these possibilities was correct. It was a bit like quantum superposition; any one and all of them could in fact be true.

Ultimately, Rusty and his team determined that despite these positive results, the Raman spectrometer wasn't stable enough and the lab needed to get hold of other, more sensitive equipment.

But one thing was certain, he said, which proved to be most unsettling of all: in terms of the timing and the effect—the dimming of the Raman spectrometer's light exactly matching the time of our collective intention—*his lab had seen results they'd never seen before with their equipment*. We may have done something to the probe or the sample of water even if we didn't know what that something was yet.

Chapter 13

Leaky Buckets

While the Penn State team was pondering how to proceed, I returned to Gary Schwartz. We thought about what had worked in our early experiments with the Germination Intention Experiment and Korotkov's Water Intention Experiments. Why not merge the two by carrying out a Water Germination Experiment? This time we'd send the "grow" instructions to the water, not directly to the seeds themselves.

There were several scientific precedents for this. Research had shown that a person's state of mind when holding water used to water plants could affect their growth. Biologist Bernard Grad had carried out a small experiment watering barley seeds with salt water, which ordinarily can stunt a plant's growth. However, he'd only watered the seeds after each vial of water had been held by one of three people: a man with a green thumb and two depressed patients. The fastest-growing plants had been watered by the green-thumbed healer, followed by those watered by one of the depressed patients, who had nevertheless grown enthusiastic about the experiment. The slowest-growing plant was watered by the vial held by the most depressed patient. Although a tiny experiment, it had a big implication, suggesting that a person's attitude could affect the water, and in turn affect anything watered by him.

After our first Water Germination Experiment, when Gary analyzed the average level of growth, the thirty seeds watered with the "grow intention" were more than a tenth of a centimeter higher than the ninety control seeds in the control water (4.77 cm versus 4.66 cm)—a 0.05-inch difference. A statistical analysis on these numbers reached borderline significance.

We did note one interesting phenomenon, though. Ordinarily, in a Germination Experiment, all the seeds don't sprout. In the latest experiment, only 90 percent of seeds in each of the control groups sprouted, but every last seed sprouted in the target group watered with the intended water. We tried it again but had to disqualify the study because of problems with the lab procedure.

We'd had some encouraging results, but for once I agreed with Gary that we should proceed slowly before the Lake Biwa event. We decided to go ahead with another preliminary experiment, which we dubbed the "Clean Water Experiment," and this time we'd go back to basics. We'd examine whether we could make changes in the way in which light rays run through water—another way of looking at whether we'd made changes in the cluster structure of the water molecules—and see if these changes could be picked up by a sensitive camera.

By then, Gary was using his GDV to examine and photograph the light patterns produced by water samples, and he'd discovered that samples varying in purity produced different patterns. Mineral water and tap water, for instance, have very different-looking GDV images. Typically, bottled water generates a larger inner "water drop," or glowing area, and a much smaller and smoother outer "aura" area than tap water, whereas tap water's image is very diffuse, like the image of the moon during an eclipse on a particularly cloudy night. As he'd already been experimenting with this, one simple possible experiment would be to ask our audience to try to transform the glow of tap water to become more like that of mineral water. This was surely an intention statement that our audience could easily engage with.

Gary's lab technician Mark prepared four petri dishes of tap water, photographed each dish, and emailed the images to me. These water samples were then allowed to sit for five days in a secured location until the date of the experiment so that all the water samples would become stagnant and their energy footprints would be similar. Water is a substance that needs to keep moving. When lake, river, or swamp water becomes stagnant and stops flowing freely, it can become a breeding ground for bacteria and organisms like pond scum, just as occurs in a polluted lake like Lake Biwa.

Mark took GDV photographs, and all four photos looked virtually identical, with an increased, diffuse aura around a blurred center, a situation that occurs when water becomes stagnant.

The two Clean Water Experiments attempting to change stagnant water worked, and the results were plain to see. All four GDV images of the four water samples had looked very similar beforehand, but after the intention, the photo of our target dish was very distinct, with a larger center water drop and a smoother aura—very much like the energy footprint of bottled water. The controls had smaller center water drops and more jagged outer auras.

We'd taken one more tiny step in showing that intention might be able to affect water, but it all felt rather theoretical to me. Before we headed for Lake Biwa, I wanted to try an experiment with some application to real life, using another measurement universally accepted by the scientific community.

The easiest way to demonstrate any shift in purification is to measure a change in pH. The pH of a liquid has to do with the concentration of hydrogen ions in water, compared to a universal standard, and it measures the sample's acidity or alkalinity. The lower any pH measure is below seven, which is neutral, the more acidic the substance is, and the higher the pH above seven, the more alkaline it is. Water pH remains fairly stable, and tiny changes of one-hundredth or even one-thousandth of a unit on the pH scale can be measured; a change of a full unit or more on the pH scale would represent an enormous shift that was unlikely to be the result of an incorrect measurement. In fact, if your body's pH goes down by just one full unit, you're probably dead. There was a precedent for using thoughts to affect pH: Stanford University physicist William Tiller had carried out an experiment attempting to change the pH in water by intention and had managed to move the pH up and down by an entire unit.

Although the plan for Lake Biwa was to attempt to raise the water's pH (the higher the alkalinity, the purer the water in most instances), we first attempted to lower the pH with a "Water into Wine" Experiment, largely because I wanted the audience to have a bit of fun—we were running the experiment a few weeks before Christmas. Participants would be asked to send intention to lower the pH of a sample of ordinary tap water so that it would be more acidic—more like wine. We ran the experiment twice, and it worked both times, although the changes were tiny and our audiences were modest—a thousand or so. Gary thin-sliced the time frame so that even the subtlest of changes would show up clearly. In this expanded scale, the pH of our target beaker was consistently lower than that of the controls, and our decrease in pH during the exact time we sent intention was paralleled by a small but measurable decrease in temperature (compared with the matched control). Something definitely was going on here, even if we were once again going against nature and trying to make a naturally alkaline substance more acidic. With our Lake Biwa Experiment the plan was to work with nature by trying to influence the water in the opposite—and more natural—direction.

At last I felt ready for the Lake Biwa Experiment. I enlisted Konstantin Korotkov, who was also to be speaking at Dr. Emoto's Water and Peace Global Forum event. We would run the experiment in person with the assembled audience, but also simultaneously online, on a special internet site, again created by Copperstrings. On Sunday, March 14, I flew with my husband; our youngest daughter, Anya, then thirteen; and her friend Helen to Tokyo. Several days later we boarded the bullet train to Kyoto, speeding past Mount Fujiyama to Kyoto and finally, via a local train, to Lake Biwa, where we were met by Dr. Emoto's family at a gala reception at Biwako Hall, the site of the conference, the following day.

Later that same evening, my husband and I climbed out onto the rocks along the shore of the choppy lake, still freezing in the March air, to scoop out two samplings of water in two different glasses to be the target and the control. We took them to Konstantin, who measured the pH of the water and then took measurements of light emissions with his GDV instrument. At the time, all measurements of both samples were virtually identical.

After Konstantin photographed both glasses, he emailed the images to me, and I sent on a photo of one of the glasses, chosen randomly by Anya and Helen, to our Copperstrings web team in India, who readied the online experiment for the following day. Once again we were creating a psychic internet: a target in Japan, a photograph held on a website in India, one audience physically present in Japan, the other virtual and scattered around the globe, all of us connected together by a tiny glass of water.

At noon Japanese time, I revealed the photo of the target water samples on my PowerPoint presentation before the live audience at the same time that our web team revealed it to the online audience on our Intention Experiment site, with the same instruction: send an intention to raise the pH of the water by imagining it as a mountain stream. I also showed them the image of a pH scale that moves from red (acidity) to blue (alkalinity), and asked the audience to move it to the right—to be more alkaline.

We didn't have to wait long to hear the results; Konstantin was able to announce them by the end of my presentation. After our intention, our target water sample had shown a rise in pH by nearly a full pH unit and a highly significant change with the GDV equipment, as compared with that of the control water. From the data, Konstantin showed a statistical difference in the signal and intensity of the light, as compared with the control glass of water. As he continued to measure the water, he discovered that the intensity of the signal carried on, suggesting that we'd made some sort of permanent change.

Rusty Roy had been intrigued by the fact that historically water has been central in important rituals. "Besides being physically necessary to life, since ancient times, water has been closely associated with the psyche, intuition, and healing," he wrote me before our joint experiment.

"Although this link has been ignored by modern medical research, most religious traditions give water a key place in their rituals—from baptisms and anointing to special blessings. It may well be that these blessings, given with true loving intentions, actually change the structure—hence the properties—of water."

We hadn't proven this with our experiment, and before we could run it again, Rusty was taken ill and died that summer. But I thought about what he was suggesting: water was used in virtually every religious tradition, not just to cleanse away impurity and sin, but also as part of a blessing. And this would mean that many religious and cultural traditions took for granted the outrageous notion, proposed by Emoto, that water can embed a thought.

I'd begun carrying out my own informal experiments with water in my Power of Eight workshop groups, intrigued by scientific evidence suggesting that water is a tape recorder.

My experiments had evolved from a demonstration conducted by Dr. Melinda Connor of the University of Arizona, which entailed asking ten members of the audience to spend a half hour in meditation, mentally sending a word (for an object like "dog") into a small empty baby-food jar filled with water. The embedders would then be asked to write the word on a piece of paper, fold it up so it wasn't visible, wrap it around their jar and secure it with a rubber band. Then I'd place these jars around the room, divide the audience into ten groups and ask them to move consecutively from jar to jar, silently attempting to intuit the object word that was "embedded" in each jar of water.

No matter where I was in the world, in workshop after workshop, at least half of my audience correctly identified at least one word of the ten "embedded" words, or something closely associated with the word (i.e., if the word was "dog," they'd get "bone").

When Peter embedded the word "barbecue" into a jar, Dorothy, standing before the bottle of water, suddenly got a strong mental image of a burger being cooked on a fire, and Sarah became very hot, as though heat were coming off the jar.

At another workshop held at a popular retreat in Austin, Texas, Janet decided to spend her meditation time in the woods near the cabin where the workshop was being held. In the midst of programming the jar with her word, she suddenly got frightened that snakes might be crawling around the woods, but knowing that the water might pick up on these thoughts, she repeatedly said to herself, "*Don't think 'snakes.'*"

When the audience was asked to intuit the word in her jar a number of people mentioned that they got the sense of something long and slimy, and a few workshop members actually identified the word "snake." In another similar situation at a retreat in Costa Rica, Annika had been sending "lion" into her jar when she happened to see a large iguana, which frightened her. Afterward, during our guessing session, Dimitri, one of the other participants, picked up "lion's mane" and a few got animals, but Diane, possibly affected by the iguana, wrote down "green alligator."

Although the weeklong Costa Rica retreat had involved just nineteen participants, we had even more extraordinary results than usual. With our first word, "conch shell," four of the nineteen wrote down "shell," Jolene wrote down "spiral," and Lissa, "funnel," and Dimitri drew a picture of a conch shell without realizing what he was drawing. For "needle," Joao picked up "needle," Nancy picked up "something sharp," Lissa wrote down "something with a point," Jolene picked up "quill," and Dimitri, "porcupine." For another jar, embedded with "tiger cat's eye marble," one saw "eye" and another a "yellow circle"; for "blue butterfly" one participant got a direct hit and Will drew the shape of a butterfly. For "crab," one picked up "fish," another, "jellyfish," Lissa saw "sharp edge," and Dimitri, "sharp nails." Out of our nineteen participants and nine jars, fourteen got at least one hit, most had more than one, and Dimitri and Kay picked up words embedded in four of the jars.

I decided to take this one step further and try it over the telephone during my teleseminars, each time embedding a word in a baby-food-size jar of water and asking the audience over the phone or on a special Facebook page to try to pick out the word.

In one phone call I held up a jar in which I'd embedded the word "banana" and asked the audience to guess the word in the jar. When I polled them afterward, one-sixth of my audience had come up with banana, a yellow fruit (several saw "lemon") or an object of the same shape:

"I saw the image of a banana, noticed a fresh banana smell, thought of banana popsicles, smelled and saw banana bread, strong banana images and smell."

"I saw a yellow, crescent moon shape."

"One of those Swiss wine flasks that are curved."

"A curved spoon sort of like a banana."

"A monkey eating a banana."

I tried it a second time with the word "star," and specifically imagined a five-pointed star. This time, one-fifth of the audience came up with the word or something with the same distinctive shape:

"I saw the image of a starfish rise up into the sky and explode into a shower of stars.

"Starfish shaped with five extensions."

"A star and cosmos or shooting stars."

"Star heart clover."

"A drawing of a star."

There are over a million words in English, three-quarters of them nouns. If you eliminate the concept nouns or those related to people, you may be left with perhaps 600,000 words. There may not be enough zeros to work out the odds of the answers I was getting being the result of simple coincidence.

The little jar experiment held many more giant implications than whether we could shift pH by a single unit. Human consciousness seems to be like a leaky bucket with our thoughts spilling out of us, getting embedded in everything from other people to our food. Remember that plants are 90 percent water and we are about 70 percent water. If we imprint certain information into water and give it to others to drink, do those thoughts affect them? Do the thoughts we have when we're preparing our food affect the people who ultimately eat it? How far can we take the tape recording in our lives?

Six of the seven latest water Intention Experiments had worked. We had offered a simple demonstration that our thoughts could change water, and from a distance, even if the changes we'd achieved were small and far less dramatic than the giant effects I'd been witnessing with the water jars in the workshops and teleseminars. But no doubt those tiny changes are remarkable in themselves. Changing the quality of water and making it alkaline by a full unit is more evidence of our enormous capacity as creators.

Even more interesting to me was what *didn't* happen during the big experiments.

I'd surveyed the participants in the large water experiments, but nothing significant had changed for them, during or afterward, the way they'd changed during the Peace Experiment. A few had cried from strong emotion and others felt a powerful connection with the other intenders, but any visualizations experienced had mostly to do with petri dishes or, at best, bodies of water around the globe. No one seemed to have undergone the kind of dramatic liftoff experienced by those who took part in the Peace Experiment. The intenders might have felt more connection with the world's water and more optimistic about the possibility of cleaning up pollution or the environment—many of our readers reported feeling at one with the water, or even experiencing a profound sensation of tasting the wine in the Water into Wine Experiment—but in every other way, nobody's life changed. Other than feeling a bit peaceful during the experiment, most people felt no long-term change in their mood, and almost no one experienced the type of mystical experiences or major epiphanies that occurred during and after the Peace Experiments. The majority claimed to have felt better on the day and gratified that they'd done something that might ultimately help the planet ("I feel hopeful that this could help with planetary clean-up in general") but promptly forgot about it. Very few experienced changes in their relationships, or changes in themselves. No one felt a sense of universal love, no one was suddenly hugging strangers, no one felt impelled to pursue a new life purpose. As one veteran from the earlier experiments wrote in, these water experiments had felt very different: "With the Peace Experiment there was a photo with children. Their eyes spoke to me."

To cause a rebound effect, I was beginning to realize, our experimental targets needed one essential element: other human beings.

Chapter 14

The Twin Towers of Peace

Rebound effects were also occurring in my Power of Eight groups. Lissa Wheeler joined a Power of Eight group to follow a dream—to write a book that would help bodywork practitioners treat patients with trauma. Lissa was not a natural writer, and by the time she joined the group, she was on her third editor and struggling with major doubts that she could ever make it happen. She was also intimidated by the prospect of marketing the book, promoting herself on social media, and getting shot down by negative reviews.

The first breakthrough came when she and her group began sending intention for Dinah, who needed support to overcome her fear of not having enough money. Dinah had even begun to talk about having to leave her home and pursue work she didn't want to do just to pay the bills.

Lissa and the other members of the group visualized Dinah receiving all the support she needed to turn her financial circumstances around. One day shortly afterward, Lissa felt a strong urge to go to a particular store, and, once inside, she happened to notice an acquaintance from long ago, who she remembered was a former book publisher. Lissa plucked up the courage to say hello and open up about her struggles with her book project. The friend offered to shepherd her through the entire process, later introducing her to a new editor and to a marketing expert. Lissa then connected with a personal-development coach to help her work through her doubts and break down the project into manageable steps. Ten months later, Lissa published her book, *Engaging Resilience*, which became a number one Amazon bestseller in two categories after its launch.

"It was like stepping onto a moving escalator of support," wrote Lissa about the rebound effect of her Power of Eight group's intention. "We didn't do that many intentions focused on my book—maybe twice. But the consistency of meeting week after week and visualizing success for whoever we were focused on built up a muscle inside me—a trust in my own possibility."

Many people in the Power of Eight circles were connecting so closely with their targets that they experienced the same effects. In one workshop in the Middle East, we sent intention to heal the arthritis Mahood was suffering in his right hip. Four of the group suffering pain from arthritis also felt markedly better after sending Mahood intention.

I wanted to test this rebound effect with another large Peace Intention Experiment and in the run-up to September 2011, there was now an obvious target. Like most Americans, I'd been forced to revisit the horror of September 11, 2001, on every anniversary for the past nine years, as every television channel relentlessly replayed the familiar sequence of events: the too-blue cloudless September sky; the first flight crashing into the North Tower, as if a catastrophic mistake; the ramming of the second plane into the South Tower seventeen minutes later, confirming there was no mistake about it; the bodies cascading out of one-hundred-story windows; the slow-motion concertina of the two towers within a half hour of each other into a cloud of fulminating black dust. It had been the we-will-never-forget imagery that America believed necessary to cling to in order to properly commemorate the dead, but with the tenth anniversary looming that summer, I grew determined to offer up an alternative.

The idea for a 9/11 Peace Intention Experiment was sparked by a chance meeting, after I'd agreed to make time to meet a friend of a friend at the Miraval retreat in Tucson, Arizona, where I was staying for a conference. Tadzik Greenberg, a genial dreadlocked fellow in his early thirties, loped over to introduce himself in the reception area. As the founder of Planet Coexist, he informed me, he'd called the meeting to discuss the ambitious plans he and his friends had for marking the tenth anniversary with a giant festival in Seattle called One: The Event, which would hold three days of global activities hoping to transform a day of fear into one of love, forgiveness, and unity. One: The Event had been planning a host of activities, speakers, and live music in Seattle at the University of Washington and the Seattle Memorial Stadium, which would also be broadcast around the world via an international webcast through scores of other peace organizations. The plan was to "shift the tides of fear and anger into love and harmony," according to the event organizer, Laura Fox, in the official press release and also to "explore what is broken in our current systems" and "what can each of us do to take action to bring our visionary solutions into the world."

Tadzik had heard about the Intention Experiment and was hoping to have the event culminate in some sort of global peace experiment. Was I interested? I stared at Tadzik's get-up, a hodgepodge of loose-fitting, patchwork garments and ancient sandals. I was highly dubious that he and his colleagues could pull off such a vast and complex event, until he rattled off an array of well-known and highly respectable transformational and activist organizations with whom One: The Event had already agreed to partner: the Shift Network, the Pachamama Alliance, Four.Years.Go., the Agape church, and many others. As the morning wore on, he began to win me over—and with good reason. Over the next few months, Tadzik would prove to be an extraordinary networker.

I spent days mulling over the kind of Intention Experiment that might best mark the date in some sort of positive way, and suddenly thought of Dr. Salah Al-Rashed. A Kuwaiti from a prominent Arab family, Salah had single-handedly pioneered the human potential movement in the Arab world. He'd been educated in the United Kingdom and the United States, and after receiving his doctorate in psychology at Eastern Michigan University, he'd returned home to set up a center and share what he'd learned from the West, offering workshops and training programs on self-development and spirituality. Salah was also a well-known peace activist, calling for peace in places like Palestine at a time when others in prominent positions like his demanded reprisal and continued conflict. In 2010, he'd started the Salam (Peace) Group, which presently had thousands of members and groups in forty Arab cities throughout the Gulf, from Gaza and Cairo to Riyadh and Abu Dhabi, each group meeting every week either in person or on the internet to send prayers for peace. Salah's own books, including a novel about enlightenment, have been massive bestsellers throughout the Gulf States. After launching his own television and radio shows, it became virtually impossible for Salah, an uncharacteristically tall and imposing bearded figure, his black hair scraped back in a short ponytail, to go anywhere in Kuwait without someone asking for his autograph. He is, for all intents and purposes, the Deepak Chopra of the Middle East, and, of all people, he would be able to drum up huge Arab participation in the experiment.

I'd met him in 2009, after he'd attended one of my workshops, when he and his wife, Sarah, who runs the center, had visited my company's London office to ask if they could host me in Kuwait the following year. We'd been impressed by the handsome couple and agreed, my husband planning to accompany me.

As it happened, Bryan couldn't spare the time off, so I had to travel alone. When I arrived at the Kuwait City airport the following February, I was momentarily panicked, frantically scanning the heavy crowd of Arab men in traditional dress for Salah, his wife, or someone holding a sign with my name on it. Finally I heard my name being called out, but the voice belonged to someone I didn't recognize. When Salah and Sarah had come to our office, they'd been dressed in Western gear, but the person calling for me was outfitted in traditional *thwab* and red-checked keffiyeh head scarf, and the woman by his side was covered head to toe in black with full hijab and niqab, her eyes the only clue to who she was.

When we got to the hotel, Salah showed me the room where I'd be speaking the following day and ran through a few cultural pointers. "The men will sit on one side and the women on the other," he said. "Don't reach out to shake a man's hand. When you do your experiential exercises, make sure to segregate men and women, and don't ask them to touch a member of the opposite sex in any way. Leave time at eleven a.m. for them to pray."

When I walked onto the stage the following day with my translator, a young woman from Syria, covered up in what appeared to be a gray-zipped coat and a tightly wrapped head scarf, we were met by a sea of black on the right side where the women had chosen to sit, naturally segregated, and white and red, where the men were sitting, on the left. The attendees of both sexes were well educated—with many doctors, lawyers, and other professionals among them—and they'd come from every nation in the Gulf, from Saudi Arabia to Palestine.

I looked out at the audience—intelligent, polite, expectant—and thought about what I was about to teach them: the power of thoughts to affect their reality. *This is going to be interesting.*

But as the first day wore on, I was overwhelmed by their enthusiastic embracing of these modern ideas about the new science and the power of intention, which they felt meshed perfectly with their religion, aspects of which were present in some form during virtually every conversation. During the midmorning breaks, the men headed to the corner of the room, got on all fours, positioned themselves in a direction so that they were facing Mecca, and leaned forward to pray. Afterward they quietly resumed their seats, even those from the most conservative countries like Saudi Arabia, happy to engage in my distinctly Western New Age ideas.

My audience was highly inquisitive, but over the two days, I became the biggest student in the room: *Why do you pray at 11 a.m.? Why do you cover up? What are you wearing underneath that black cloak?* (Answer: Gucci.) *Does covering up lower the incidence of rape? What do you think about not being allowed to drive? How would you solve the Arab-Israeli conflict?* The outpouring of love from all of them in response to my curiosity was extraordinary, a gratitude for the attempt at being understood. In the end, I had to buy a suitcase to hold all the presents they showered me with: framed photographs of one attendee with her arm slung around me; elaborate silver and turquoise jewelry; models of traditional ships of Kuwait, which used to be a major port; religious artifacts, including souvenirs from the Kaaba, located in the middle of Mecca, the holiest site in all of Islam.

Salah went on to host a few more of my workshops in Dubai and Turkey, and I'd loved the audience every time I'd gone. His followers were the perfect group to provide my Western audience with a counterpoint to al-Qaeda.

Over the summer, he and I fleshed out our plan to call for a new Twin Towers of East and West in communion and solidarity for peace. It was Salah's idea to open the event by apologizing on behalf of all Arabs, but I told him the West needed to apologize as well. However justified America felt in invading Afghanistan after the 9/11 attacks, the fact remained that the Afghans had lost far more than we had. Most Westerners did not acknowledge that some one hundred thousand innocent Afghan people had been killed, injured, detained, or deported because of a war set off by a small group of Arab radicals, who were terrorizing them as well. Peacemakers such as my friend James O'Dea, former head of the Washington bureau of Amnesty International, who'd witnessed public trials in such war-torn areas as Rwanda, convinced me that one of the fastest routes to restoring accord is frank and public apology for past wrongdoing.

When we started to plan our 9/11 Peace Intention Experiment, I was careful to use a design that was identical to that of the 2008 Peace Intention Experiment. We would repeat our intention every day for eight days, as we had in 2008, and keep the main members of our original scientific team: Gary, Roger, Jessica Utts.

Salah and I were of one mind about the target: it had to be Afghanistan. By the time of our experiment, the war had been raging on for almost ten years. The Helmand and Kandahar provinces of Afghanistan, the two large provinces in the south and the major strongholds of the Taliban, had incurred the highest number of war- and terrorist-related injuries and deaths among both military and civilians of any province in the country. Both areas had been the sites of recent car bombings and suicide bombers, and, as the largest opium market in the world, this area of Afghanistan, which borders on Pakistan, was also the target of terrorist attacks from outsiders. Those involved in fighting the NATO forces' "War on Terror" were a mix of Taliban fighters and warring tribal groups involved in the opium trade. After the 2010 peace initiative attempted by the then Afghan government with the Taliban broke down, and after NATO had initiated new offensives as a consequence, the violence had intensified.

Copperstrings designed a web platform that was virtually identical to the one we had used for our Peace Intention Experiment in 2008, but with two differences: we'd have two sets of the same web pages, one in English and the other in Arabic, and we'd rent an even larger server power, as a double insurance against our site crashing. Since One: The Event intended to broadcast the entire three-day event over the internet, Tadzik put us in touch with a woman who had just started up an internet TV station and who offered to do a daily livestream of Salah and me after each day's experiment by hooking our broadcast into the webcast of the event.

For the broadcast, Salah started out with an unabashed apology on behalf of all Arabs for not being more vigilant and for allowing the attacks to happen, and I returned the apology for the West's "aggressive and violent response to 9/11" and offered a pledge "to work to avoid violence and political and economic exploitation by offering an alternative to war and to Western economic and political supremacy at any cost." Both of us also promised to "work for greater tolerance of differences of all faiths and creeds."

When it was time for the intention to begin, the web pages flipped over again, this time to reveal an image of an Afghan boy surrounded by white doves and an image of Caucasian and Arabic hands clasped—a symbol of the East and West coming together.

This time our 9/11 Peace Intention Experiment attracted participants from seventy-five countries, from Iceland to Brazil to California to Indonesia, and also every Arab country on the planet. People participated in all sorts of ingenious ways: via the giant screen at One: The Event, from a mountaintop, during a Native American peace pipe ceremony. One participant who was driving during the time of the experiment pulled over each day to participate. "I could always feel an energy shift about ten minutes past the hour when the experiment began," he said. In the end, many thousands participated from One: The Event and the simultaneous broadcasts, with seven thousand more signed up on our website and tens of thousands tuning in to my daily webcast. This was undoubtedly the biggest mind-over-matter experiment in history.

On the third day of the Peace Intention Experiment I was encouraged after discovering that the United States had endorsed plans for a Qatar-based office, located in Doha, for the Taliban to begin proposed peace talks with the West. But after the experiment ended on September 18, once again we had to embark on a patient three-and-a-half-month wait, to allow events to unfold over the rest of 2011 so that we could determine whether our intention had any effects, while I had to find somebody inside the American military willing to disclose the true figures to me.

No official in an American conflict wants to talk about the bodies. I spent several months hounding officials inside virtually every large agency involved in the War on Terror: the US State Department; UNAMA (the United Nations Assistance Mission in Afghanistan), which tallied civilian casualties; the Afghanistan government; the combined-forces mission inside Afghanistan and various departments inside NATO, which ultimately referred me to the International Security Assistance Force (ISAF), a NATO-led mission set up by the UN Security Council initially to train the Afghan Forces, but whose powers had grown to leading the combat operations in the regions.

Most of the agencies would not release all of their figures—ISAF claimed to have no tabulated information they were prepared to make public about military casualties, but they had plenty of data about the enemy's attacks and civilian casualties. UNAMA had monthly data about individual sections of the country for 2009 and 2010—but not for 2011.

After making a nuisance of myself with the ISAF, I was finally put through to its official spokesman, a German general named Carsten Jacobson, who was a bit more helpful although cagey about releasing any data. He warned me that any statistics on military fatalities aren't completely reliable, because once a soldier is wounded, he usually is transferred back to his home country, and NATO's combined army usually doesn't get any further feedback about whether he has lived or died. Largely to get rid of me, he eventually sent me an official report from NATO's Afghan Mission Network Combined Information Data Network Exchange database, about the war's progress over several years as of January 13, 2012, and I was also finally able to get hold of UNAMA's 2011 annual report concerning civilian casualties. Both sets of figures may have represented a sanitized version of casualties, but as I was using them for the whole of my comparison, at least they'd be consistent.

Both reports had largely done the work for us: a comparison of casualties among military and civilians as well as various kinds of enemy attacks with those of prior years, and a complex analysis of trends in various kinds of violence, so that we did not need a professor of statistics to produce the final numbers. The statistics in the reports included the numbers of enemy-initiated attacks in different sections of Afghanistan, including the south, the target of our intention, and also the number of improvised explosive devices used, including mine strikes, the principal means by which Afghan insurgents execute strikes against the NATO military and the cause of more than 60 percent of civilian casualties, according to ISAF. From all these figures, we could analyze what happened in September 2011 and the two months afterward as compared to what had happened in the months and years prior to our experiment.

Once again, we were amazed by the huge drop in the casualty rate that occurred among civilians and the military after the 9/11 Peace Intention Experiment, specifically in our two provinces. For civilian casualties, according to the NATO statistics, 440 civilians were killed in August 2011, but the monthly numbers fell to 340 in September and continued falling over October (290) and November (201), representing a 22 percent, 14 percent, and 30 percent drop over the months before, respectively. All three figures were well below the average death rate (374) that had occurred the 28 months prior, with October 2011 23 percent lower than average and November 46 percent lower than average. In fact, November 2011 represented the second-largest percentage decrease of civilian casualties since the beginning of 2009. Overall, between September and November 2011, civilian casualties fell by an average of 37 percent, compared with the casualty rate in August 2011.

In terms of enemy attacks, NATO figures show that attacks with explosive devices fell by 19 percent, remained at the same figure in

October, and continued to fall another 9 percent in November and 21 percent more in December. This final figure was 16 percent lower than the average attack rate from September 2009 to December 2011, the two-plus years before.

Perhaps the most interesting downward trend had to do with overall initiated attacks by the Taliban. The monthly figures for 2010 had showed a steady upward trend of attacks (up 80 percent overall in 2010), but then this trend flattened out and hardly changed until the beginning of 2011, when attacks began climbing relentlessly upward until August. After our experiment in September, the numbers began a steep downward trend, falling drastically from October to December 2011, and overall enemy-initiated attacks over the last three months of 2011 were 12 percent lower compared to the same period in 2010. As the report noted about the second half of the year: "This is the longest sustained downward trend in enemy-initiated attacks recorded by ISAF."

In fact, compared with the rest of the country, the southwest—the target of our intention—recorded the largest decrease in figures compared to the September of the year before, an extraordinary 790 percent decrease over the month before and a 29 percent decrease for the entire year compared to 2010. This trend carried in October (the attack rate was 500 percent lower), November (400 percent lower), and December (300 percent lower).

What made our results even more compelling was the fact that the big decreases in violence that had occurred in our target provinces of Helmand and Kandahar had not been uniformly experienced around the country. The overall figures of casualties for the entire country bounced upward in December 2011 after two suicide attacks during the Ashura celebrations in Kabul and Mazar-e-Sharif, and Taliban-initiated attacks increased by 19 percent in the East from 2010 to 2011.

But, once again, what did this all mean? As with 2008, nothing definitive. *You construct a hypothesis, and when it pans out, you have to test it again. You test it again, it pans out, and you have to try it a few more times. Only after results are replicated four, five, six times can you show a pattern that starts to get interesting.* And, once again, there were a million and one circumstances that could have accounted for the decreases in violence. For one thing, there was the fact that the United States and NATO had already begun to wind down the Afghan war, although that did not explain the concentrated lowering of violence in our two regions. Despite many potential variables, the results did seem compelling, particularly considering that we had quantified our intention request—as we had in our 2008 Peace Intention Experiment for Sri Lanka—asking that violence be lowered by at least 10 percent. When we looked at the data for the whole of the country, it consistently showed average decreases in casualty rates somewhere around the 10 percent mark.

Besides a straightforward analysis of casualties, I'd also asked Roger Nelson to see if there was a demonstration of any effect on the Global Consciousness Projects' network of random event generators during the eight days of our collective intention, just as he had done for our 2008 experiment. Nelson linked together the eight days of data to make a sequence that included all the machines' output during all the twenty-minute time periods of the eight days, paying particular attention to each day's ten-minute windows of actual intention. After the third day, he found a very steady trend—a general tendency for the outputs that accumulate during each second of the time period we were looking at to be similar.

"Most of the deviations are negative," he wrote me, which meant that the mean was less than the expected one hundred, like tossing a coin and having it constantly come up tails. When Roger strung together the deviations, the graph line showed a continuous downward direction. "A persistent or 'steady' trend reflects consistency," he wrote me, "and that in turn suggests an effect that isn't just chance."

Roger cautioned me that the effect size was very small compared to inherent "noise"—or chance data. "Deviations which appear in our graphic displays are a combination of possible effects and ordinary random fluctuation," he wrote me. One single experiment like this one cannot be reliably interpreted on its own.

But when he compared his results with those of the 2008 Peace Intention Experiment, he discovered a virtually identical negative trend in the cumulative deviation graph. "This similarity across the two experiments helps support an interpretation of the negative deviations shown in the current data set as an effect linked to the Intention," he wrote me.

Something definitely seemed to be happening there, as it had with our original Sri Lankan Peace Experiment, but something else was happening that I began to notice on Facebook, Instant Messenger, and the two surveys I'd conducted of the participants about their experience, one in English and the other in Arabic. We appeared to be ending the war in another way.

Chapter 15

Healing Wounds

From the first day of the 9/11 Peace Intention Experiment, the participants had made an extraordinary connection with one another, even more powerful than during the 2008 Sri Lankan Experiment—in most cases the most extraordinary connection they'd ever experienced.

"Like I was a piece of metal being drawn to a magnet not of this world from my elbows to the tip of my fingers," wrote Logan from Switzerland.

"Like I had a white glow around my body, with a white cylinder connecting my body (and everyone else's) to the target area," wrote Cathy.

"Like being in the total vortex of prayer energy of all who were focusing, like an out-of-body experience," wrote Linda from the United States.

"Like swimming in an ocean of goodwill, love, hope," wrote Simona from Romania.

Their bodies felt "electrified," and many were shaking "like when you're really cold and you get 'the chills,'" with "waves and waves of shivers all over" their bodies. They became cognizant of internal sounds, as though "there were people whispering" in their minds.

Many were openly sobbing during and after the experiment, as though they'd "tapped into a global pain body" intensifying their own feelings. "I wasn't a body at that very (long) moment," wrote Saad. After reading aloud the intention, Michel's throat was so sore that he had to stop speaking. "It's the closest," wrote one, that "I have ever felt to 'God.' "

Just before the experiment started, Logan had texted his sister to ask if she could get to a computer and sent her the link, even though she is not a practicing meditator and had never tried to do intention. After the experiment, she called to tell him she'd gotten so emotional during the experiment that her partner even wondered whether she'd been looking at an upsetting photograph because she'd been crying so much. "I told her that is exactly what I felt," he said.

They'd hallucinated strange, highly specific utopian visions from their own perspective, feeling as though they were "IN their bodies, but also right OVER THERE in the target area" in Afghanistan:

"A white energy of peace shooting from all of us, mingling into a wide beam of light and hope!" wrote Amal.

"People working together to rebuild schools, hospitals, and lives and a country of love and peace!!!" wrote Debbie.

"Afghanistan as the very source for the new Global Peace in the world," wrote Cornelia.

"The kids running beside the rivers . . . heard the birds singing and saw schools and universities in Kandahar and Helmand . . . then I saw the West and the East normally mingling together, no difference at all," Fatima wrote.

"The white birds of peace getting out of Ground Zero covering the world," Tarik wrote.

"All the rancor in Washington DC and US politics dissolving like chocolate," wrote Maridee.

"George Bush with Condoleezza Rice and [Donald] Rumsfeld living and sitting between all Afghanistan people having a drink with each other, like friends," wrote Marjorie.

"Arabs and Americans . . . all throwing their arms down into a huge crater, all working to cover it all with earth and then placing a sign upon it that reads, 'Here lies War, gone forever,' " wrote Linda.

As the week carried on, thousands continued to tune into the web TV station I'd teamed up with to do a daily live stream update on the event. During the daily broadcasts, which had an instant messenger chat room, many of our Western participants began to instant message and befriend people from the Arab countries who could write in English—and vice versa. The resentment and suspicion about Arabs was beginning to transform into love and acceptance. The Westerners began wishing the Arabs well—*"Ante diemen fee kalbi"* ("You are always in my Heart")—and as they began to feel connected to the Arabs, "like a support from the right side one can virtually lean on, like feeling brothers from far away," their attitudes toward the Middle East began to shift: "forever will Afghanistan be synonymous with Peace for me." The pain of 9/11 and lingering rancor was healing.

"The experience of IM'ing with people from Egypt, Saudi Arabia, and many other Middle Eastern countries—during the IM messages, we wished each other peace and expressed love—made me cry," wrote John from Tucson. "It was very therapeutic for me—a citizen of the USA."

As word of our experiment got around, it began to create some positive effects, even among those who had not participated. May Lynn attended her book club during the week of the 9/11 Peace Intention Experiment. Her friends were commenting about how overloaded they felt with negative 9/11 emotions. "I was able to tell them that there is a large group of people working to use this anniversary to improve peace in Afghanistan," she wrote, "and they were really glad to know that this experiment is happening!"

Samuel from New York, who teaches a large population of Middle Eastern students, told them about our experiment. "They were quite surprised and now want to continue," he wrote.

Many of the Arab participants reached out in friendship to the

West: "We are brothers, we will always be here for you. Although I don't know you, I feel a connection with your pure souls."

"This day is the day that we all felt the loss and no one felt the gain," wrote Bahareh. "Your God is my God. My God is your God."

Following Salah's lead, the Arabs started apologizing to the Americans, a point of view that "millions of Arabs and Muslims are sharing with him."

"In six minutes," said Kholood, "he said what I'm trying to say in years."

They began bringing the apology to their own lives. One of my respondents, feeling challenged by some people who didn't agree with him, apologized for not sharing their point of view. "It's suddenly okay with them," he wrote. "Was it the apology?"

Both sides began discussing ideas on Facebook about how to create peace between East and West: "Stop using the words 'East' or 'West,'" "change it from East and West to World," "call it WEast."

As with the 2008 Peace Experiment, participating in this experiment had brought peace into their lives, particularly their relationships. Three-quarters of my participants spoke of how their newfound sense of peace had improved relationships in every regard:

"Family relationships."

"My neighbors."

"My sisters."

"My twin brother."

"My dogs."

They were getting along better with clients, ex-husbands, siblings, neighbors, those they normally argued with, even employers. "My husband came to me midweek, and said that I was more approachable and open. The little stuff did not bother me." Many made a pact with themselves to resolve lingering conflicts with others and heal rifts, even with those who'd caused them pain. Saad let go of the "negative energy" he had toward a friend and forgave him. "On the first day, I was holding hands with a friend I just made peace with, after a long time of not talking to one another," said Susan from Spokane. "We held hands throughout the experiment, and when we were done, we hugged."

A third of the participants were getting along better with people they normally dislike or argue with. An ongoing conflict with a husband was brought to "full confrontation, but then moved quickly into resolution and solutions." Arguments over an accident, landlords, sisters-in-law that had carried on for years suddenly got resolved. One found it difficult to agree with the commercial attitudes of his coworkers or to go along with directions from the managers he disagreed with, but found it "easier to love them." Others were able to tolerate people they didn't usually get along with: "I felt compassion for my not-so-nice boss."

"I kept flipping between the target area images and the 'warlike' energies emanating from my next-door neighbor," wrote Stephen from New Orleans. "I felt that the Peace Intention Experiment would heal BOTH situations."

Those from both East and West had experienced a powerful opening of the heart, and once again, a majority were falling in love with everyone they came in contact with. They experienced a "more peaceful feeling toward everyone," "an openheartedness that continued between the intention meditations." They'd become "more open and comfortable and at ease around people" and "less concerned about what they think," "experienced more clarity and kindness toward personal issues," feeling "compassion and empathy toward others increasing." They felt a "'gentling' of mind-set," "more tuned-in," their hearts "more open in general," and more willing to "let stuff go."

Many had completely transformed in the way they related to other people. They felt able to see "people and situations more clearly," noticing when they were judgmental of others and themselves. They found anger "more uncomfortable than before," were "more apt to apologize and forgive," had "stopped reminding themselves" of what the other did to hurt them, and "now were not taking things so personally." They felt a certain "urgency to let go of the past hurts," were "feeling feelings more," were "listening more without judgment," and were more desirous of sharing their personal truth.

"I see myself in everyone I meet, experiencing their feelings, finding compassion."

"Recognized my need to extend my Love to ALL Humanity."

"More connected to strangers and the world community."

"More compassion for all people."

"More open to make contact with strangers."

And once more, these positive effects seemed to spill over to other areas of their lives. Many claimed to have had "personal miracles" happen in their life, experienced "the most creative periods in the last five years," had a "spiritual quantum leap," enabling them to be more intuitive and more sensitive to others, and seen a "big improvement in healing skills," in the case of one therapist. "My life," wrote Abdul, "has changed for the most beautiful."

They'd been loath to leave the pure love of the circle and end the experiment, but once they had, many felt hope for their country and the rest of the world and a greater urgency to be an instrument of change, "an overwhelming need to continue to concentrate efforts in Helmand and Kandahar" or "to make a tangible contribution to other areas of the world like Rwanda, Congo, and other places on the African continent."

"I must find other people here who want to do this on an ongoing basis," wrote Martin.

"I felt," wrote Rose, "that I am a part of the solution."

In a way, the 9/11 Peace Intention Experiment had been a giant exercise in multicultural prayer. Mohandas Gandhi, who believed that all religions "were as dear as one's close relatives," advocated the power of different faiths praying together:

> . . . religion does not mean sectarianism. It means a belief in ordered moral government of the universe. . . . This religion transcends Hinduism, Islam, Christianity, etc. . . . It harmonizes them and gives them reality.

A two-year national study published in 2014 by researchers at the American Sociological Association found that community groups across America that embrace multi-faith members, such as Christians, Jews, and Muslims, find praying together a "bridging cultural practice."

"We aren't talking about superficial team-building exercises," said University of Connecticut professor of sociology Ruth Braunstein, who studied the phenomenon. "These are practices that are central to groups' cultures and emerge over time as participants reflect on the qualities that unite everyone in the group and develop shared rituals that are meaningful to everyone."

On May 3, 2015, NewGround, an interfaith organization that focuses on strengthening the bonds between Muslims and Jews, organized an event they called Two Faiths One Prayer, to gather Muslims and Jews in common prayer. They started with some twenty people of the two faiths praying together on a Los Angeles beach, gathering up more and more of the faithful from both religions throughout the day as they traveled together on public transportation and moved to five other locations. The group was a hundred strong by the time they'd reached a rooftop in downtown LA for their evening dinner, with Muslims reciting their nighttime *Isha*, and Jews reciting liturgical poetry, or piyyutim, at Los Angeles City Hall.

"It was kind of like an aha moment," said participant Maryam Saleemi. "We're praying to the same God, why aren't we doing this all the time together?"

But even with these "bridging" efforts, no one had examined the rebound power of collective praying, its ability to heal the personal wounds of the healers themselves.

Ellen, one of our participants, found the exercise could heal her lingering grief over the loss of two friends. During the experiment, she wrote, she could not stop weeping. A good friend of hers, Lee Shapiro, and his soundman, Jim Lindelof, had been killed in Afghanistan in 1987 while making a documentary. Their bodies were never recovered. "I kept seeing an image of them. The energy seemed so huge," she wrote. "This was a profound experience for me."

Toni's life had been torn apart when her sister and children had been murdered by the children's father just a few weeks before 9/11. In her eyes, the Peace Experiment saved her life. "Change happened that for a second destroyed all my faith until the love of the community and signs from the universe restored it and made me more grateful than ever," she wrote. "I poured a more intense Love than anyone else out into the universe that day as my heart shattered and soared simultaneously. The world remembered while we mourned. And many lives were changed forever."

I had no idea if my experiment could take the credit for the improved peace in those two southern provinces of Afghanistan. But if the feedback from participants were anything to go by, the act of sending intention had created peace in their hearts that seemed to be transforming their lives and views of East and West. For many on either side, the experience had been extraordinarily healing, a simple means of breaching ideological divides.

The outcome of the actual experiment again was almost irrelevant; the real healing was happening with the participants. Joint prayer had itself brought the East and West together, had proved to be profoundly uplifting, and had given hope to many on both sides.

"Thank you, world," Yasser wrote. "You're still a good place, with all these peaceful people."

I didn't know if God had answered our prayer for peace, but certainly our prayers had given us a glimpse of God—and even a fleeting glimpse of heaven on earth. "I had the sense that although we had a specific 'target,'" said Aimee, "we were healing everyone everywhere at once."

Chapter 16

The Mirror Effect

Ingrid Pettersson's husband died in late 2013 only four weeks after being diagnosed with a rare cancer. Although his oncologist had been confident that his cancer was treatable, he was deeply affected by the pessimistic attitude of his home-health nurses and their gloomy prognosis, particularly their repeated pronouncements that he'd never resume normal activity like driving again. Ingrid stood by watching helplessly as her husband seemed just to give up.

As a result of his rapid decline and death, Ingrid had to close his thriving business and move out of their new apartment in Gothenburg, Sweden. Within months she was beset with financial difficulties. For most of the early part of that year she was overwhelmed by shock, grief, and depression at her dramatic loss and her suddenly changed circumstances.

Four years before her husband died, she had joined one of my workshops and experienced a profound transformation in a Power of Eight group. Her whole appearance changed: her skin was shining, she'd felt much younger and energized than she used to be, and was healthier than ever. "I felt so good and I attracted more of what I wanted in life and even my relationship improved," she said. Her friends and even her doctor remarked upon this major difference in her health and appearance, a change that continued on for about a half year, but as she returned to "old habits" all the good changes slowly "dropped away."

Several months after her husband's death, she remembered that experience and decided to join in with a large-scale experiment of ours targeting a person with posttraumatic stress disorder. After participating, her debilitating grief vanished. "Since your last experiment, it is all gone," she wrote. "I could not believe it. It is just amazing." For the first time in months Ingrid had a good night's sleep and woke up energized and felt happier than she had in a long time. "The negativities and even my grief after my husband died did not seem to affect me as much as it had done during the last months." And best of all, she said, "I got back the flow in my life." After the experiment she decided to start a new career organizing workshops in energy healing in Gothenburg and Stockholm.

Ingrid gave me another important clue about the rebound effects of the global Intention Experiments. Her epiphany had occurred during the first global Intention Experiment that targeted a single human being in early 2014. Up to that point, I'd allowed only smaller groups—Power of Eight groups or Intentions of the Week—to target people. I'd avoided any formal large-scale experiments on human subjects mainly because I was unsure whether a group of intenders numbering in the many thousands would have a positive or negative result, particularly after the increased violence that had occurred during the Sri Lankan Experiment. After the Intentions of the Week had some success, the Peace Experiments turned out to have positive results, and many of the global experiments suggested that size of group made no difference to the outcome, I decided to press ahead with our first big experiment on a human being. This was going to require the most careful baby step of all, and an opportunity to test it out had fallen into my lap in October 2013, when I was invited to deliver a few lectures in Hawaii.

Along Bishop Street, amid the modern glass and steel high-rises of downtown Honolulu, stands a bit of architectural whimsy, the Dillingham Transportation Building, a particularly fine example of Italian Renaissance Revival and a monument to Hawaii's most famous double act, Benjamin Franklin Dillingham and his son "Uncle" Walter, both of whom recognized that the key to transforming this sleepy little clutch of islands into a modern cash cow was sugarcane and a means of moving it from one side of the islands to the other. The father built the railroads, and the son, through his own construction company and a number of political favors, drained the wetlands, extended a number of ports, and finished the job of commercializing the islands. Two floors above the ground-floor arcade and stone quoins of the Dillingham building's exterior and the gilded Art Deco lobby in a small corner suite resides the office of another father-and-son team with an equally immodest goal: to change the face of modern medicine, using state-of-the-art video technology.

Like the Dillinghams, the Drouins are transplants, French Canadians from Quebec, Dr. Paul Drouin, a medical doctor who practiced for twenty-five years by integrating the best of conventional and alternative medicine. The censure of some peers, his growing frustration about the closemindedness of the profession with its total unwillingness to entertain the worth of any kind of alternative medicine, and the implications of new discoveries about quantum effects in biology led him to a big idea: to create a university that would give doctors and health professionals an opportunity to learn about the new science and alternative theories of treatment and to incorporate this knowledge into their careers.

Dr. Drouin's vision began to take shape after he joined forces with his then twenty-five-year-old son Alexi, who holds a degree in film and television, and who had his own big idea: to make the university entirely virtual. He would film leading authors, academics, and practitioners lecturing about quantum physics or their work in modalities of alternative medicine in front of a green screen and embed these courses on iPads, which would be automatically offered to every student. Thanks to Alexi's perfectionism and technical virtuosity, the green screen transformed into a modern television news desk, and the lectures were professionally presented, complete with PowerPoint slides.

By then the duo had moved to Honolulu, where the process of getting accredited held less red tape, and christened their fledgling institution the Quantum University for Integrative Medicine. To date, the university has enrolled nine thousand students, many of whom have gone on to complete their PhDs. "I'm the form, he's the content," says Alexi, pointing to Dr. Paul, as the students know him, an irrepressible sixty-five-year-old with a thick French accent and a jack-o'-lantern grin, who regularly features on many courses as the well-loved face of Quantum U.

Students and teachers have a single opportunity every year to connect in person with each other at the university's annual conference, and it was there, in October 2013, where I was invited to speak that I met the Drouins and began to kick around the possibility of doing an Intention Experiment with them on their web platform. One night during dinner with the Drouins and the other conference speakers, Dr. Jeffrey Fannin, director of the Center for Cognitive Enhancement (now known as Thought Genius) generously offered both to donate his time for the entire project and to find some willing volunteers to participate in the experiment. Dr. Fannin holds a PhD in psychology, with a particular interest in neuroscience, and has a good deal of experience in studying and, through EEGs, mapping the brain waves of mental disorders such as anxiety, depression, or attention-deficit hyperactivity disorder. As Quantum University has its own television station, the Drouins could broadcast the event on web TV, Alexi assuring me that they had more than adequate bandwidth for the thousands we expected to sign up. Ever the technical innovator, he also had a plan to increase what my British neighbors refer to as the bread-and-circuses aspect of the event: he was fairly sure that he could show the effects on our subject's brains happening in real time.

In the next few months while we were planning how to achieve such a complex technical feat, two patients of Dr. Fannin suffering from anxiety offered to allow themselves essentially to be experimented upon using what is, by any regard, a most unusual therapy: the power of strangers' thoughts. One would act as the target; the other, the control, but both would be blinded to who had been chosen as the subject. During the months prior to the experiment, Alexi helped to publicize the event by sending out numerous Facebook announcements, and by the time we were ready to go, we had more than seven thousand people signed up.

Alexi had set himself up for his greatest technical challenge to date. On April 24, the day of the event, his cameras would show alternating split screens between Jeffrey, who'd be connected via Skype; me on another Skype screen; Dr. Paul, who would be moderating from the studio; and both patients, who'd be attached to EEG machines. Although he wouldn't feature much in the broadcast, an EEG was also attached to Mario, our "intender" sitting in another room, who would participate with our audience in sending intention to the chosen target so that we could compare his brain waves during intention with those of our target.

Human brain waves come in different speeds, from the very slowest, which are delta and theta (5–8 Hz or cycles per second), associated with deep meditation and sleep, to alpha brain waves (of 8–13 Hz), which also occur during light dreaming or meditation, to beta waves (around 13–30 Hz), for everyday cognitive tasks, and gamma (above

30 Hz), a state of extreme focus. Jeffrey's work entails translating the results of EEG readings into a tomography, or qEEG, showing different frequencies of a person's brain waves and comparing them to "normal" brain waves, and his equipment can display in real time the percentage of certain brain waves present in a person at any moment. For our broadcast, Alexi had also rigged up an additional screen to be trained on both EEG machines, showing the percentage of different brain waves currently activated, which were depicted as flashing bands of different colors, stretching and receding horizontally.

On the day of the event, thanks to Alexi's skill, all the screens meshed together beautifully. Our choice of the target, chosen from a top hat, turned out to be Todd Voss, a veteran of two wars, the Persian Gulf War and the Iraqi war, who'd been diagnosed with PTSD after coming home. He suffered a deep depression and was hypervigilant. Whenever he went into a room, he found it necessary to sit with his back to the wall, always scanning and looking for threats. He also had trouble sleeping. The Veteran Association's response to his condition was to prescribe a load of medications, but Todd knew the drugs would just put a "Band-Aid" on his experience, and he was eager for any nondrug solution to his problem. Our intention was to attempt to calm down Todd by at least 25 percent and also to focus on increasing the alpha waves of his brain (the brain waves associated with greater calm and peace).

With brain mapping, a person's entire brain frequency activity is depicted as a "map" of thirty little "heads" in different colors, each representing certain frequencies of brain waves. Green depicts wave frequencies that correspond most to "normal"; and a rainbow of other colors are used to show how much a person's brain waves deviate from normal (red, for instance, shows several deviations more than normal; blue, several deviations lower than normal).

Jeffrey had carried out brain maps on both Todd and Kathy, our control person, before the experiment and would be doing so again during the experiment, after the experiment, and then a few weeks later, in mid-May.

We had asked our audience to concentrate on increasing the number of Todd's alpha waves, which were depicted as bands of turquoise, by having them stretch out and become more prominent; during the experiment, courtesy of the split screen, we were fascinated to watch the effects of our collective intention in real time as the bands of turquoise began to stretch.

Brain mapping done before the experiment had revealed certain areas of Todd's brain with a frequency "signature" characteristic of PTSD, but various brain maps made during the Intention Experiment determined that Todd's alpha waves had increased to three standard deviations above normal after our intention. Most exciting of all, the area of the brain most representative of PTSD was almost completely normal during the experiment.

Other analysis demonstrated that coherence within the brain—the ability of the brain waves to work better together and stay working together—also had improved. After Dr. Fannin worked out what is called an independent t-test to determine the statistical significance of the experiment, he discovered less than a 1 percent probability that these results occurred through chance.

The same effects were not evident in the brain maps of either Kathy or Mario, our intender, both of whom experienced virtually no change in alpha brain waves. This appeared to rule out the possibility that the changed outcome might be the result of the placebo effect, particularly as neither Kathy nor Todd knew whom we'd chosen until after the event.

The results were initially very encouraging, but there were a few problems with the study design we had to acknowledge. One difficulty with an experiment of this type, which uses a highly novel medical intervention, is finding willing volunteers and also carrying out such an experiment at reasonable cost.

For potential targets, Jeffrey was limited to those among his own patient base who were willing to undergo such an experiment, most of whom had already begun treatment with him. Todd Voss had previously undergone two kinds of brain training, one with Dr. Fannin, and part of this training involves teaching techniques to lower stress by increasing a person's own alpha brain waves. However, when Todd's symptoms returned, Dr. Fannin regarded him as a deserving candidate for our experiment.

A week after the event, Todd described himself as very much improved—well enough to plan some extensive travel. He no longer felt the need for any more clinical sessions for his PTSD, and within the next year, would go on to get married and have a child. His clinical experience and recent brain maps were compelling, but because Todd was previously taught techniques that purport to achieve the exact effect that we were attempting to achieve by intention, it was impossible for us to declare categorically that any changes in his brain were due to our thoughts alone, rather than his own brain training.

While it was gratifying to me that Todd was feeling so much better, an even more compelling part of the story emerged when I checked in with our participants several weeks later. This time, nearly one-fifth reported some sort of pronounced physical improvement.

"My carpal tunnel injury improved, and I felt very relaxed. Even slept better."

"I suffered with my knee for almost three years. After this experiment all the pain I used to have was gone, completely."

"A previously chronic condition in my back and knees is feeling better."

"Last ten days I have regular digestion (I had constipation for almost twenty years)."

"I had relief in my hip as if I had taken some pain relief."

"The pain in the knee is completely gone."

"Painful hip issue appears to be healing."

"I believe my body is 'recalibrating' in some way."

"I used to have colon problems, not anymore. :)"

"Continuing improvement to skin condition."

"No longer experiencing sciatic pain."

"I sleep better, and anxiety and panic attacks have disappeared."

"Having suffered from rheumatoid arthritis for the last few years, . . . I am seeing subtle, but regular, signs of improvement. I am having less pain and anxiety."

"I feel like I am finally ready to deal with my own PTSD on a physical level."

The results were even more amazing in the weeks that followed. Nearly half of our respondents were healing their relationships: clients, ex-husbands, siblings, neighbors, parents. This time the emphasis was not simply on more peace in relationships, but on healing old wounds. Sandra reconnected with her mother, whom she ordinarily only spoke to by phone a handful of times a year. "We had a conversation as we have never had in my life."

Two participants reconnected with their sisters, forgave past hurts, and were able to see each other "with new eyes." "I'm getting along with my older sister and that never happens. It's like her heart is softening or opening," said one. Another participant healed a relationship with a coworker. Marie healed her connection with her husband. "My husband looks at me like I met him yesterday and it feels so good!"

"My life—everything about it—my health, relationships, outlook, energy level, happiness, openness, etc.—just keep improving," wrote Sophie. "I've plainly shifted."

I thought of Ingrid Pettersson's grief, and it finally dawned on me what might be happening: the rebound effect on participants mirrored the intention itself. If they prayed for peace, their lives became more peaceful. If they tried to heal someone else, they experienced a healing in their own lives. Focusing on healing someone else brings on a mirrored healing.

Chapter 17

Going in Circles

The same mirroring was occurring in our Power of Eight circles. In a workshop of mine in Maarssen, Netherlands, Bet had sprained her ankle and her arm after a fall and was chosen by her group of eleven to be the recipient of the healing. Once she joined the circle, she realized that she was acting as both sender and receiver at the same moment. "When I felt people putting their hands on me I thought, maybe I'll join them because after all it is me. I felt the energy coming in and I decided to be the eleventh person to join in the ten others, and then I felt my energy being part of a whole energy."

Bet was no longer a single entity. She was both the sender and the receiver at the same time.

It was now clear to me that the particular rebound effect experienced by our participants had to do with the object of focus, and that this was causing the mirrored healing. In my book *The Bond*, I write about the discovery by Italian neuroscientist Giacomo Rizzolati that when we observe an action or an emotion in someone else, in order to understand it, the very same neurons fire in us as though we were performing the action or having the emotion. He'd called these copycat brain waves "mirror neurons," but what seemed to be occurring among my participants was something beyond simple mirroring. They weren't just reflecting back. They were identifying so strongly with the object of their intention that they seemed to be merging with it, as though it were happening to them:

"I . . . tasted blood and smelled blood as though I were in Afghanistan and had lost family this way."

"Todd Voss, I hear the name constantly in my mind. It's as if he has become part of me."

What does mirroring to this extent do to your brain, I wondered. Does it cause something to change permanently? Richard J. Davidson, a psychologist at of the Laboratory for Affective Neuroscience at the University of Wisconsin–Madison and his colleague Antoine Lutz, a research scientist at the French National Institute of Health and Medical Research, are fascinated by the workings of extreme brains—those that have had an unusual, lifelong workout, particularly from long bouts of meditation—and how their neural connections and structure continue to revise through life, depending upon the focus of thought.

"A brain region that controls the movement of a violinist's fingers becomes progressively larger with mastery of the instrument. A similar process appears to happen when we meditate," Davidson and Lutz wrote in *Scientific American*.

Working with monks and Buddhists connected with the Dalai Lama, Lutz and Davidson have studied which parts of the brain change, whether it's focused attention, where the meditator concentrates on the in and out breaths; mindfulness, where participants maintain moment-to-moment awareness of all their sensations, including thoughts, to develop a less reactive response to them; or loving-kindness meditation, where the meditator focuses on a feeling of compassionate love toward all other people.

Each type of meditation offers a workout for a different portion of the brain and at different frequencies: focused attention and loving-kindness meditation appear to activate very fast frequencies in the brain (beta 2 at 20–30 Hz and gamma waves at 30–50 Hz), which tend to create a brain more accustomed to intense focus, whereas mindfulness makes use of very slow brain waves (theta brain waves of 5–8 Hz) so that the brain relaxes and becomes less concerned about reacting to what's going on in its environment.

The object of focus tends to improve the brain in remarkable ways. Focused attention and mindfulness create meditators with heightened perceptual awareness: focused meditators to the outer life and mindfulness to the inner life. As I discovered when I wrote about this in *The Intention Experiment*, when the object is compassionate meditation and a desire to send love to all things, these types of thoughts send the brain soaring into a supercharged state of heightened perception. The brain begins working at fever pitch; Davidson's studies of monks showed their brains produced sustained bursts of high-frequency gamma waves, or rapid cycles of 25–70 cycles per second. This is the kind of brain speed experienced only when the brain is at rapt attention, when trying to dig through working memory for something, during great flashes of insight. At this speed, the brain waves also begin synchronizing throughout the brain, a state necessary to achieve heightened awareness, and the two sides of the brain also begin operating more synergistically.

As Davidson's research with monks demonstrated, achieving a high gamma state activates the left anterior portion of the brain, the portion associated with joy, and exercising this "happy" part of the brain seems to produce permanent emotional improvement. Brain states achieved in these circumstances can create such positive effects on people's emotions that they stay that way permanently.

Our Intention Experiments and Power of Eight circles appeared to make use of a combination of focused attention and loving-kindness, as we had both an altruistic goal—to heal something—and a specific focus—either a specific person or a situation, such as a war-torn region of the world. It may have been possible that my participants had undergone some sort of a "happy brain" effect as a result of their compassionate intention, which is usually only experienced by monks who have spent years at compassionate meditation.

But could this have been the X factor causing the major changes in people's lives?

Tania Singer, director of the Social Neuroscience Department at the Max Planck Institute for Human Cognitive and Brain Sciences in Leipzig, Germany, runs the ReSource Project, a large-scale study in which participants are taught a protocol of Eastern and Western mental training devised by Matthieu Ricard, a French Buddhist monk and cellular biologist and a close colleague of the Dalai Lama's. After eleven months, the group is then studied to see whether this training changes anything in their lives and relationships.

After just a single week of carrying out compassionate meditation, Singer's participants become more cooperative and more willing to help others in need, as demonstrated by having them play a virtual "prosocial" game designed to test and measure their desire and capacity to help others. Singer's meditators became more desirous of helping, even if there were no prospect of their good deed being returned, and more sensitized to distress signals from people—all marks of an enhanced sense of wanting to connect with others.

This certainly accorded with the experiences of my participants in the Peace Intention Experiments, a majority of whom felt a greater readiness to get along with others, including strangers, but they hadn't required nearly a year's worth of training—just a single ten-minute session of sending intention.

Neuroscientist Mario Beauregard refers to gamma brain states as "oceanic," enabling you to move out of your small self to something larger, and recent experiments of his suggest that these gamma frequencies are infectious. He embeds hidden gamma frequencies into music, which he plays to some volunteers during a meditation, then measures their brain frequencies. Repeatedly the brains of his listeners begin to evidence a "resonance response," a greater percentage of these same high frequencies. Just listening to these hidden frequencies, even if you don't know you're doing so, trains your brain to imitate them.

"If you change the brain waves," he told me, "you change a person's sense of identity. The sense of self shifts and the self becomes larger."

When experiencing this oceanic state, they move from the small self, which is solely interested in their own reactions to the exterior world, to a larger self. "In that expanded state, they can more easily let go of chronic emotional patterns and limiting beliefs," he says. "And it is easier for them to feel universal love."

One of his participants was a mother, age sixty-five, whose son had killed himself when he was seventeen. The event had occurred twenty-five years before, but she was still suffering from guilt and trauma over the manner of her son's death. After experiencing the gamma brain wave state, says Mario, she felt finally free of her grief and ready to move on with her life.

So in our circles, if the individuals' brains had flashes of a happy-brain synchrony, this may have helped to increase a natural sense of joy. But the healing effects my participants were experiencing in their lives and relationships seemed to go beyond an oceanic feeling of universal love. What was it about the circles that had this effect? Robert Cialdini, a psychologist formerly at the University of Arizona and author of *Influence: The Psychology of Persuasion*, maintains that a sense of connectedness increases altruism: people have a natural desire to help when they lose their individuality and step temporarily into a state of oneness.

Lutz and Davidson discovered that practicing compassionate

meditation produces more activity in certain parts of the brain called the temporoparietal junction, the medial prefontal cortex, and the superior temporal sulcus, all of which are usually activated when we are altruistically inspired to help someone else. Altruistic thoughts of healing also appear to change a number of networks in the brain—the orbitofrontal cortex, the ventral striatum, and the anterior cingulate cortex—all associated with compassion, positive emotions, even maternal love.

I could see why people could heal their relationships and feel more contented with their lives through this experience, but what was it about the experience that enabled them to heal their own physical conditions?

We know from extensive research that meditation has direct effects on the body, helping to diminish inflammation and alter the functioning of important enzymes, including an enzyme called telomerase, which affects the longevity of a cell. In Singer's own work at the ReSource Project, her meditators enjoyed improved immune system and nervous system responses and a lowering of stress hormones.

But there was something more. In my experiments and groups, these healing effects were instantaneous and recorded largely in people not practiced in meditative techniques. Love must get bigger through our Power of Eight and Intention Experiment circles, and this creates a virtuous circle of healing. It may well have been that within the confines of intention groups, it finally becomes safe to give, and giving might ultimately be the entire point of the exercise, the aspect of the intention that proves to be the greatest healer.

It was time to turn the experiment on its head. Instead of studying the outcome—the getting—I needed to study the process: the act of giving.

Chapter 18

Giving Rebounded

George seemed to be a lost cause. He had been diagnosed with a low-grade glioma, a deadly type of brain tumor, and as a university-employed biomedical researcher, he was well aware of the prognosis: no cure; slow but steady growth over months; and—this, the outside possibility—certain death within two years, no matter what the treatment. Surgery was out of the question and chemotherapy or radiation, largely ineffective.

With no decent prospects afforded by conventional medicine, George began looking for a miracle, which led him to the first North American Pentecostal Church conference. There, a ministry team prayed for him, and George felt the Holy Spirit surge through him, a giant heat and a vibration like an electric shock that caused him to fall and cry out. He was so overcome that he followed Pentecostal Church conferences around every region of North America like a dedicated groupie, each time standing in line to put himself forward for healing and each time, after the ministry teams prayed for him, experiencing the same lightning-bolt reaction. But every three months, when he returned to his medical doctor for an MRI scan, hoping for some evidence of a change of outcome, there was no discernible effect on his tumor, which continued to grow.

By 2004, still optimistic that prayer would work for him, George joined a ministry trip to Cuba run by the Global Awakening Movement, the ministry arm of the Vineyard Church Pentecostal revival set up by Randy Clark, a minister of a Saint Louis, Missouri, church. Clark had been the unwitting architect of the so-called Toronto Blessing revivals in 1994, which began with one meeting of 160 people. When Clark invited the Holy Spirit to come, the church attendees suddenly shook, exploded with giggles, and fell down as though in a drunken stupor, and many claimed miraculous physical or emotional healings. Word quickly spread, and tens of thousands more from around the world flooded into Toronto to attend the nightly prayer healing sessions, which Clark carried on with for twelve years. By the time the revival had run its course, some three million people claimed to have felt its extraordinary effects.

As a member of one of the ministry teams in Cuba, George joined the group praying for a man with severe vision problems who could barely see more than two feet in front of him. After fifteen minutes of prayer, the man claimed to be able to see objects twenty feet away without glasses. Another one of their subjects was a woman who'd been so ill and emaciated from ovarian cancer that she'd been unable to eat or walk, and during the prayer she experienced the same reactions George had, but after falling down, she regained her strength and ability to walk. Her tumor, which had been easily palpable, could no longer be felt.

George was so impressed by this experience that after returning home, he began praying for other people at every opportunity, sometimes devoting entire evenings and weekends to his healing attempts. He also continued to attend Pentecostal conferences in the United States and signed up for more traveling with Global Awakening and other ministries to Latin America and Africa, but increasingly, his focus turned to praying for others. When he could no longer handle the sheer number of people asking for his healing, he created a prayer team, which remained behind at least an hour after every Sunday service to pray for the long line of people waiting their turn.

After two years, George's doctor noticed that George's tumor stopped growing—in fact, it was starting to shrink, and he no longer had any of his earlier symptoms. During the following eight years, he remained symptom-free; in fact, his most recent scan even omitted the word "tumor" from the report, a fact confirmed by Candy Gunther Brown, an associate professor of religious studies at Indiana University who documented cases like George's for her book *Testing Prayer*.

Although his healing had not been as immediate or dramatic as so many of those he'd claimed to have witnessed, both George and Randy Clark, who wrote up his story in his book *Changed in a Moment*, attributed the turnaround in George's health to the cumulative effect of every prayer that had been said for him, like a savings account that suddenly amasses a critical mass of interest.

But as I had come to see it, it wasn't the amount of prayer George received, or the fact that he'd been prayed for at a batch of conferences or missionary trips. George began to get better the moment he began praying for someone besides himself.

After the Healing Intention Experiment, I began to consider that another powerful force might be responsible for all the miracles that I hadn't accounted for: the rebound power of praying for other people.

Dr. Sean O'Laoire, an Irish Catholic priest with a PhD in transpersonal psychology, had stumbled across rebound prayer when looking for something else. O'Laoire was especially curious about the effect of prayer on emotional and mental health, an area largely ignored by scientific research, and as a practicing priest and clinical psychologist, he was uniquely placed to study it.

O'Laoire planned to focus on whether those being prayed for experience any changes in psychological states like anxiety or depression and mood, and his advertisement for volunteers in the Bay Area newspapers produced 406 volunteers. The 90 who'd put themselves forward as willing to do the praying received training sessions, which included some intention and visualization techniques, just as I do in my workshops.

There was no question that prayer had a positive effect on the participants. All 406 of O'Laoire's participants improved on every objective and subjective measure of physical and psychological health. But when O'Laoire looked a bit closer, he saw that those doing the praying were doing even better than their targets. And although the amount of prayer hadn't made a difference to those being prayed for, it did impact those doing the praying. The more prayer they carried out, the healthier they became.

O'Laoire was shocked by his results. "It seems, then, as if praying is more effective than being prayed for," he concluded. The bank account of prayer was just the opposite of that imagined by Robert Clark. The compounding of your interest in the bank account entirely depended on the amount you prayed for others.

So if the healing effects experienced by my participants were some sort of by-product of altruism, what would cause the rebound effect? Karl Pillemer of Cornell University was interested in just this question of altruism in relation to healing—interested enough to devote a big chunk of his life to a single study. He enlisted nearly seven thousand older Americans, many of whom had volunteered to help with projects attempting to address environmental issues such as pollution or toxic waste, and many others who assiduously avoided this kind of volunteer work. Pillemer tracked the health histories of all seven thousand of his subjects for twenty years, and he wasn't disappointed. At the end of this study, he discovered that his volunteers were far healthier and physically active and half as likely to be depressed as the others.

Offering your time to work for the greater good obviously produces more than just a warm and fuzzy feeling; it proves strengthening to both mind and body. In fact, there are health-giving effects in simply focusing on anyone besides yourself. As had been the case with George, if you're suffering from some sort of condition, you're more likely to overcome it once you turn your attention to someone else. That was the conclusion of one study of more than eight hundred people suffering from severe stress who were followed by University of Buffalo researchers for five years to compare the state of their health with the extent to which they'd helped anyone outside the home, including relatives, friends, or neighbors.

That little bit of helping acted like a bulletproof vest. When faced with future stressful situations like illness, financial difficulties, job loss, or death in the family, those who'd helped others during the previous year were far less likely to die than those who hadn't. In fact, the contrast between the people who'd helped and those who didn't could not have been starker. When faced with each new stressful event, those who'd decided not to lend a hand increased their chances of dying by a whopping 30 percent.

As Father O'Laoire had discovered, directing your attention to someone else is particularly good for your mental health, and the reverse is also true; people who are strongly focused on themselves are more likely to suffer from depression and anxiety and negative mood. In fact, if you have to choose between giving and receiving, there is no longer any question that it is better for your health to do the giving: in one study of older Americans, those who gave experienced less illness than those who were on the receiving end of their kindness. And of all the religious coping behaviors relating to better mental health, one of the major ones among a group of mentally ill patients was giving religious help to others.

Giving also appears to be central to longevity. Research from Stanford University in California of senior residents showed that those who were involved in volunteer work, particularly in religious groups, had mortality rates nearly two-thirds lower than those who weren't—a situation, noted the researchers, "only partly explained by health habits, physical functioning, religious attendance and social support."

Something about the desire to do something for someone else with no strings attached or personal benefit has an impact on health and well-being far and above that of anything else: diet, lifestyle, social support, or religious belief. Of any single lifestyle factor, altruism appears to be the ultimate vitamin pill for assuring a long and healthy life.

The act of giving also has a huge effect on happiness. After the political scientist Robert Putnam of Harvard University wrote his groundbreaking book *Bowling Alone*, which woke Americans up to the fraying of the social fabric across the United States, researchers at the John F. Kennedy School at Harvard decided to explore exactly what makes for what they refer to as "social capital"—happiness, close-knit communities, and satisfied residents—by carrying out a survey of thirty thousand members of diverse communities across America.

What they found was revelatory. Unless you were poor, money just didn't do it for people. Once you achieved an annual income above seventy-five thousand dollars, your emotional happiness had very little to do with your bank balance. People below that income were miserable because they were struggling just to pay the bills, but once they'd achieved that level of income, making any more money didn't offer any greater joy. That division—between being able to pay your bills and not being able to—was the only watermark connecting money in any way with life satisfaction.

But the one factor that did make for the greatest sense of satisfaction and happiness was lending a helping hand. In fact, those willing to give of their time or money were 42 percent more likely to be happy than those who weren't.

There's a physical component to the joy that people experience when helping others, which psychologists refer to as a "helper's high." When people who volunteer are surveyed, they frequently describe feeling physically the same as they do when they engage in vigorous physical exercise or a bout of meditation: the body releases endorphins, those feel-good brain chemicals that counter all the biological effects of stress. Although researchers argue that this sort of altruistic behavior requires direct contact with others, our experiments showed that it works just as well virtually, and just a photograph is enough to make the connection.

So altruism was good for your health and happiness, perhaps even the most important insurance plan for longevity, but did it also have anything to do with why my participants experienced such profound changes in their relationships, particularly with strangers? To figure this out, I needed to study something fundamental about what happens to people when they put themselves second. I needed, in short, to know a little more about the biology of goodness.

Chapter 19

Thoughts for the Other

I'd presented thirteen of my Power of Eight groups with a most unusual challenge, particularly as they were focused on personal improvement: get off of yourself. Stop doing any intending for anyone inside your group, including yourself, and focus on the other. I had some particular "other" in mind. On December 8, 2015, Luke, then fourteen, had broken up with his first proper girlfriend, and in an extravagant gesture of adolescent existential angst, had thrown himself off a forty-foot structure onto hard ground.

Miraculously, Luke survived, but everything in his body was broken. After his fall, he had sustained fractures to his skull, eye socket, pelvis, ankle, and one heel and one elbow. Besides all these broken bones, he had a brain injury, a punctured right lung, double vision in one eye, nerve damage in his pelvis, a major chest infection, a spinal injury at the bottom of his spine, and urinary tract nerve damage. He'd gone through several emergency operations and had now also to recover from the trauma of surgery to his pelvis, spine, and elbow. All this was in addition to his suicidal state of mind. He was fighting for his life but still lacked the will to live. His stepfather, Michael, contacted me to ask if we could do an intention for him.

This case spoke to me possibly more than any I'd ever received. I still had a teenage daughter at home at the time. It could have been her or any one of her friends. I made Luke an Intention of the Week and a focus of Power of Eight groups on Sunday, January 10, while Michael kept a weekly running commentary on Luke's moment-by-moment progress.

Right after our intention Sunday evening, Luke was quite agitated—which his parents attributed to an aftereffect of the group healing. A day later, as his chest infection started clearing up, the nurses could stop the antibiotics and he began to get better sleep. He was more conscious and began asking more questions about his recovery and trying to visualize himself back home and at school.

I asked my audience to send another intention to Luke the following Sunday, and Michael sent through another detailed progress report. Luke's brain injury had stabilized. His elbow fracture was healing and he could now use his arm for weight bearing, as he could with his left ankle. All his infections had cleared up and he was able to come off painkillers and antibiotics. The double vision in his left eye had cleared up. He was allowed to start moving from bed to wheelchair and his bowels had improved. Suddenly he became impatient to get better and to get back into the gym.

Michael sent though a picture of Luke whizzing around in his wheelchair, exploring the ward and hospital—"a major achievement!"—and then another one of him high-fiving the camera with his three best friends who'd come to visit him.

Everything was going in the right direction, when Michael wrote back with another update that week: Luke still didn't have feeling in his bowel and bladder, which was causing him a great deal of discomfort and pain. He'd had to continue to use a urine bag and there was a threat of a colostomy bag too. His mood had crashed. On more than a few occasions he remarked that he didn't want to live anymore. Later that day, Michael wrote to confirm that Luke was not in a "good place emotionally."

We ran a third intention on January 24 and got a note from Michael a few days later.

"Luke's mind and body have responded massively to the healing intention on Sunday 6 p.m. In fact, his mother noticed an incredible improvement in his state of mind at exactly 6 p.m., right at the time of our intention," he wrote.

At that very moment he said, Luke's mood transformed. He began speaking more positively with his mother and shortly thereafter he cleared his blocked intestines, after which he was no longer in pain. For the last four days he'd refused to get out of bed; right after the intention, he was eager to go to the physio gym. He had positive chats with his mother about returning to school and going on a family summer holiday. "This is incredible progress considering that he was having suicidal thoughts again only last Friday!" wrote Michael.

The following week, Luke was due to meet with his doctors, and with all this improvement, it was likely he would be given the all clear to start walking. But the most startling change of all was a positive change in his mood. His mother had noticed a very "confident and positive" Luke. "This is a massive step forward!" wrote Michael.

Later that week after doctors fitted him with a different catheter system for his bladder, Luke no longer had a urine bag to deal with, which improved his spirits and well-being further. His mental state was now steady, even hopeful. He could see himself back at school after the summer break. He didn't mind repeating the year.

Had that just been a case of good doctoring, or had we somehow willed Luke to want to live? Both Michael and his wife, Clair, were convinced that this change in Luke was a direct result of the healing intention. "The improvements were so sudden and totally unexpected," wrote Michael. I could say only one thing about this complete turnaround in Luke's situation: it wasn't a placebo effect. Luke did not believe one bit in what we were doing. Like most teenagers, he thought his parents' belief in the power of healing intention was stupid.

Andy was one of those who'd kept up the healing vigil for Luke, and once she started focusing on the other, things started shifting in her life.

Over the last six months, she'd tried everything to remove old patterns that had interfered with her ability to make a good living. When she joined the group, she shared with them a very specific intention she wished to put out about finding a dream job that would also provide ample income: "I am easily and joyfully allowing $20,000 or more a month net to flow to me in many joyful ways," she set as her intention. "I am being highly compensated for doing the work I love which feels like play to me." Her other intention was "trying to get clear on my new business, speaking, teaching, and healing" after closing her retail gift store of eighteen years in 2013.

None of intentions the group tried were working for her. She'd even attempted, as I suggested, to go back to her "seed moment" when she first experienced limiting thoughts about herself, and imagine changing that situation in some way. Andy remembered one very specific moment as a child when she had a very visceral feeling about just not having enough money. That still didn't bring about any big change in her fortunes.

She then started experimenting with altruism—shifting her intention to Luke and others outside her group, and suddenly she got the breakthrough she needed. "Two days after that, I got an unexpected offer to do product development and strategy for an online organization involved in human development, a job that would joyfully bring me money doing work I love, which feels like play to me!" said Andy. When she realized this might not be an ideal fit, she had the courage to draw "boundaries" with the CEO, turn it down, and accept a job coaching for a well-known and highly respected coach "with great integrity and alignment with my values."

"At times, focusing on one's own intention may be the metaphysical equivalent of a watched pot not boiling," Andy later wrote me. "Focusing on the good of others and being of service takes the focus off ourselves in a way that allows movement without noticing the passage of time. Perhaps altruism is the secret way of both consciously and non-consciously NOT observing so the desired outcome can occur."

Dacher Keltner, a psychologist at University of California at Berkeley, has made it his life's work to counter the prevailing view that human beings are hardwired to be selfish for one simple truism: we're healthier and happier in every way when we do the giving and not the taking. In his book *Born to Be Good*, Keltner quotes Confucius about the cultivation of *jen*, a Chinese concept which means that he who wishes to establish his own "character" does so by bringing "the good things of others to completion." *Jen* is essentially a rebound effect: your global well-being, your essential nature, in fact, is defined by how much you do to help others to flourish.

To figure out how this works, Keltner asks us to consider the function of the vagus nerve, one of the longest of the body, which originates at the top of the spinal cord and works its way through the heart, the lungs, the muscles of the face, the liver, and the digestive organs. Keltner notes that it has three functions: to connect with all the communication systems involved with caretaking; to slow down your heart rate, calming the effects of any fight-or-flight autonomic nervous system activity, the body's response to stress of any sort; and to initiate the release of oxytocin, a neuropeptide that plays a role in love, trust, intimacy, and devotion. If oxytocin is the "love hormone," as it is generally referred to, the vagus nerve, maintains Keltner, is the love nerve, a hypothesis given weight by the work of Chris Oveis, one of Keltner's UC Berkeley graduate students. Chris wanted to determine whether or not activation of the love nerve helps to nurture universal love in a person and a greater acceptance of differences between the self and the other, and to do so, he set up a unique type of research project involving a batch of his fellow university students.

During the study, Chris showed one group of student participants photos of malnourished children—the ultimate of the world's victims. As soon as the students saw the photos, their vagus nerves went into high gear. The same effect was not produced in another set of students, who were shown photos designed to elicit be-true-to-your-school pride, such as images of landmarks on campus or U Cal sporting events.

But the most interesting effect occurred when the students were shown photos of twenty other groups of strangers who were markedly different from them: Democrats, Republicans, saints, convicted felons, terrorists, the homeless, even students from their strong competitor, Stanford University. Those students love-bombed by their own vagus nerve reported feeling a far greater sense of similarity to all the disparate groups than those who'd been exposed to photos designed to elicit pride. Activity of the vagus nerve helped to remove a boundary of separation, causing the students to focus more on similarities rather than differences, and those feelings of similarities increased, the more intensely their vagus nerves fired. Even students identifying themselves as Democrats suddenly recognized the similarities between themselves and Republicans.

A closer look at the results revealed something even more fascinating: this group of students felt the greatest sense of common humanity to all those in need—the homeless, the ill, the elderly—whereas those whose sense of pride had been activated identified themselves far more with the strongest and most affluent groups, such as lawyers or other private university students. Instead of identifying with the people most like us, when the vagus nerve is fired, we are prompted to feel closer to *the other*, particularly the people in need of our help—and more prone to reach out to them.

Research at Stanford University discovered similar effects in a group of volunteers being trained in a simple Buddhist loving-kindness meditation. First they were told to imagine two loved ones standing on either side of them and sending their love, and then to redirect these feelings of love and compassion toward the photograph of a stranger. After that simple exercise, as a battery of tests revealed, the meditators experienced a greater willingness to connect with strangers, compared with group given a similar exercise but without the loving-kindness meditation. Just a simple statement about expressing love for all living things activates the love nerve and prompts a person to put those statements into action in the world.

This might explain why my audience felt far more open to strangers after participating in the experiments, and why so many of the Arab and Western 9/11 Peace Intention Experiment participants began forgiving each other. The compassion elicited in my participants by the Peace and Healing Intention Experiments may have activated a nervous system complex that created a greater willingness to connect with the "enemy"—indeed, all of humanity.

In biological terms, activating the vagus nerve and increasing levels of oxytocin, as occurs when we show kindness or compassion toward others, has a marked healing effect on the body. David Hamilton, the former medical researcher and author of *Why Kindness Is Good for You*, made a study of the healing effects of increased oxytocin levels and found evidence that they lower inflammation and boost the immune system, aid digestion, lower blood pressure, heal wounds faster, and even repair damage to the heart after a heart attack.

Oxytocin is so protective that it can defend us against bacterial assault. In one groundbreaking Austrian study carried out at the medical University of Vienna, ten healthy men were first injected with disease-causing bacteria on its own and then given the bacteria plus oxytocin. When first injected with the bacteria, the men displayed evidence of rapidly increasing levels of pro-inflammatory cytokines—evidence of increased inflammation. However, these cytokines were reduced markedly when oxytocin was injected at the same time. Oxytocin even plays a key role in turning undifferentiated stem cells into mature cells, which also help in repair and renewal.

The pure act of no-strings-attached giving, so scarcely experienced in our modern society, may also prove healing, as François Gauthier from the University of Quebec discovered when studying three examples of healing at the Burning Man festival. Held in Black Rock Desert in Nevada every summer, the festival runs entirely on a transaction-less system. Other than the space and toilets, the first aid, and the burning man effigy, the organization provides nothing. Attendees must provide their own food, water, and shelter, and other than an entrance fee, no money is allowed to exchange hands, which encourages an elaborate system of barter and gifting, including healing. The "gift" itself of healing someone else and "the primacy of social relationships" proved to be the most powerful therapy for many of those who came to the desert to be healed of psychic or physical wounds. "When Burners give of themselves and work in [the] healing and well-being of others, they work together for their own well-being and their own healing," Gauthier noted.

I was still puzzled about the healing effects of the Intention Experiments and why they would overcome long-standing conditions until I came across perhaps the most compelling piece of research of all about the transformational effects of altruism. It had been carried out by psychologists at the University of North Carolina at Chapel Hill who wanted to examine the difference in likely future health between healthy people who live a fulfilling life of pleasure—what we'd normally define as the good life—compared to those who live a life of purpose or meaning.

The researchers examined the gene expressions and psychological states of eighty healthy volunteers in both groups. Although the members of the two groups had many emotional similarities and all claimed to be highly content and not depressed, their gene expression profile couldn't have been more divergent. Among the pleasure seekers, the psychologists were amazed to discover high levels of inflammation, considered a marker for degenerative illnesses, and lower levels of gene expression involved in antibody synthesis, the body's response to outside attack. If you hadn't known their histories, you would have concluded that these were the gene profiles of people exposed to a great deal of adversity or in the midst of difficult life crises: a low socioeconomic status, social isolation, diagnosis with a life-threatening disease, a recent bereavement. These people were all perfect candidates for a heart attack, Alzheimer's disease, even cancer. In a few years, they would be dropping like flies.

Those whose lives were not as affluent or stress-free but were purposeful and filled with meaning, on the other hand, had low inflammatory markers and a down regulation of stress-related gene expression, both indicative of rude good health. If you have to choose one path over the other, the researchers concluded, choosing a life of meaning over one just chasing pleasure is undeniably better for your health.

This all sounds counterintuitive to us in the West, with our emphasis on material success at any cost, but it has to do with what exactly constitutes "meaning" in our lives, and the best way to gauge that is what ultimately helps ill people get better—the one aspect of life that will turn around a serious illness. Scientists from Boston College discovered this when trying to figure out why patients suffering from chronic pain and depression markedly improved in both disability and mood once they began helping others in the same boat.

As they repeatedly noted to the researchers, it was all about "making a connection" and being provided with "a sense of purpose." Our need to help other people is perhaps the one element that gives our life the greatest meaning.

But there was another factor here not covered by any of this research. The altruistic acts of my participants have been performed within a giant virtual prayer group, and this appeared to offer some sort of amplification of healing power. Certainly there is a long tradition throughout all the sacred writings of all the major religions describing the healing effects of spiritual belief and practice in a group. Those who regularly assemble in churches to pray together have been shown to have lower blood pressure, enjoy far stronger immune systems, spend far fewer days in the hospital, and are a third less likely to die, even when all other factors are controlled for. Scientists believe that those who are now age twenty who never go to church can expect to live seven years less than those who attend more than once a week.

It isn't just someone's religious fervor or assembling as a community; the collective spiritual practice appears to be as important as the group effect. One study found that those living in a religious kibbutz and praying together had about half the mortality rate of those in a nonreligious kibbutz. Like the groups who put meaning at the forefront of their lives, those who attend church frequently have stronger immune systems than those who attend less frequently, as measured by lower plasma interleukin-t (IL-6) levels. An elevated IL level is a marker of one of the degenerative diseases, such as Alzheimer's, diabetes, osteoporosis, or AIDS.

Religious belief on its own is strengthening, but not as powerful, it seems, as the group experience of prayer.

In fact, the collective aspect of the praying may be the essential factor in creating the healing effects. Rich Deem, who was healed of an incurable disease in 1985, went on to study prayer and healing, particularly the evangelistic meetings in rural Mozambique. Prayer leaders recruited blind and deaf subjects from the meetings for healing touch, and doctors were on hand to measure auditory and visual acuity before and immediately after the prayer. All were in prayer meetings involving groups of people, and all involved touch. Of ten subjects, all but one improved in visual acuity and also all but one had their hearing improve. The scientists then compared these results with those achieved by studies using suggestion and hypnosis on subjects with the same maladies. Although those studies achieved statistical significance, they paled in comparison to the results achieved by collective prayer.

It is clear that altruism brings out all the loftier emotions in us; it might be the emotion that most defines our humanity—our sense of a life well lived—and gives our life a sense of meaning. It may even be the key to whether we live or die. But the powerfully transformational mechanisms at work in my healing intention groups appeared to be the unique power of group prayer coupled with a deliberate focus away from the self.

It had been there all along, in the early Christian teachings, all those homilies so familiar that they now sound like words on a Hallmark card: *Do unto others. Love your neighbor as yourself.* Focusing on someone else heals the healer.

All this research was leading me toward a heretical thought. Maybe the endpoint of the "I want, I get" good-life scenario is that it ultimately kills you. *I want, I get—I get sick.* The key to a long and healthy life is living a life that concerns itself with a meaning beyond satisfying the needs of number one. I considered how dangerous some tenets of the "self-help" movement might ultimately be. All that focus on the self could ultimately be terrible for your health, and highly unnecessary. The quickest route to rewriting your own life's script was simply reaching out to someone else. And if that is true, the entire New Age premise of intention—using the universe as essentially a restaurant with you the customer ordering whatever dinner you happen to fancy—was wrong.

Getting what you wanted in your own life started with the readiness to give.

As my husband once wrote, Jean-Paul Sartre was mistaken. Hell is not other people. Hell is thinking there are other people. Bryan was talking about the fact we are simply a single consciousness and the falsity of thinking that we are separate. I would just add a little coda. In seeing yourself in the other, in joining together as one, other people, it turns out—particularly a small group of them praying with you—are your salvation.

Chapter 20

A Year of Intention

Patty Rutledge, an attractive fifty-five-year-old with titian hair, had challenges in every area of her life save one: her marriage. Her husband, Stephen, and she had recognized each other as soul mates from the moment they met and knew within six weeks that they would marry. Stephen was a widower with four teenagers, all traumatized by their mother's sudden death from brain cancer, and came with a great deal of baggage. Each of the children presented certain challenges for Patty as a stepmother, as did her relationship with her own very demanding stepmother, who'd married Patty's father just a year after her mother's death. Patty was very close with her four grandchildren but yearned to be closer to her three stepdaughters and stepson. Patty also had serious health issues. She'd suffered from Epstein-Barr virus and severe chronic fatigue on and off for thirteen years. "I've continued to live life and even travel, yet most of the time, I feel pretty exhausted," she said.

And there were 250 others, ordinary people like you and me, whose lives weren't working quite as well as they would have liked, and who, in the hopes of shifting something, had agreed to be part of the most radical experiment of all: a teleseminar class that would divide the audience into Power of Eight groups who would work together virtually for an entire year. Up until that time I had only tested the effects of the Power of Eight over a weekend, and the groups had almost exclusively focused on physical healing, but I had begun to wonder whether working in a group for a full year would change the group members in other ways as well. *Would everything in their lives begin to heal?*

I'd announced this "masterclass" teleseminar starting in early 2015, which would start with a training course for several months, after which we'd organize the participants into small groups of eight or so and track their monthly progress over the rest of the twelve months. This would put the power of the groups to the ultimate test with a long-term experiment out in the field. Here would be a giant petri dish of my very own, to minutely observe month after month over an entire year.

We gave each group a Greek name—Triton, Chronos, Helios, Proteus, Morpheus—set them up on Google Hangouts or Skype, and encouraged them to meet at least once a week. Every Friday I threw out new challenges to them via email, and every ten weeks I'd have a conference call with them to answer questions and check their progress further. The groups were to take turns intending for each other and then switching to outside targets, and to fill out monthly forms to see if these intentions correlated in any way with any major changes in their health, relationships, career, finances, or life's purpose. I emphasized that they were not to invent any effects.

Within a few weeks of setting up the circles and getting familiar with the technology, most groups had bonded closely; within a few months, I began to get feedback of healing effects similar to what had begun to happen to Patty.

At the time Patty started the course, she had given up spending time in the gym. In the past, after exercising, she would have to nap or accept feeling horrible all day—"like I used up the battery charge in that one hour," she said; the best she could manage was walking the dog ten minutes twice a day. Although trained as a life coach and counselor, she couldn't work much because of the fatigue. To make matters worse, an MRI scan using a contrast medium had recently confirmed that she had two lumps in her breast. Although tests showed the lumps weren't cancerous, thermography results showed carcinogenic potential.

Medical tests had confirmed that she had heavy metal overload, which was wreaking havoc with her endocrine system, and evidence of the presence of *Stachybotrys chartarum*, or black mold, 150 times greater than normal. For years, Patty had been diligently trying out various forms of alternative medicine but hadn't made much headway in getting her old energy levels back.

Besides her own health, Patty also worried about Stephen, who had a propensity to put on weight and difficulty sticking to a healthy diet. A Western-trained doctor, Stephen remained open-minded about Patty's integrative approach to her health but gave it no credence.

At the beginning of the course, Patty had a tall list of very specific requests for her Triton group: to improve her health and energy levels; forge a deeper relationship with her stepdaughters and stepson; inspire Stephen to lose forty pounds, to transform his diet, and to work out regularly with a coach; and achieve more clarity on how to best use her professional skills. But when the breast lumps showed up, that became the priority. In August—month four—she asked her group to focus on finding and healing the source of her fatigue and having the two breast lumps just melt away.

At first, her health situation did not shift in any way. After a detox regimen over the summer, she'd seen a 99 percent drop in her heavy metal levels and in the black mold but no improvement in her energy level. If anything, it worsened. Two trips for her niece's wedding in

Utah and a trip to Washington, DC, to see another niece left her exhausted.

With the group's support, Patty launched into a batch of alternative initiatives—a liver cleanse, juicing, Qigong, acupuncture, and biofeedback—all the while performing regular intentions and visualizations in which she imagined her blow-dryer melting away the two breast lumps. On August 26, she had her first breakthrough: a repeat breast scan revealed that the lumps were completely gone.

Despite this positive development, her energy levels remained unchanged and felt even more depleted after a trip to Santa Fe, when she discovered that she could hardly walk up the stairs. "I was ready to bawl," she said. "My legs kept feeling like wet noodles." When she returned home to Virginia, she began researching and discovered that she might have been dehydrated in Santa Fe's increased altitude, which would have affected her liver stores of glycogen, the energy that gives her muscles the power to move. This was a lightbulb moment. Tests by her naturopathic doctor revealed that her hunches were correct: her liver wasn't producing or releasing glycogen properly.

That same month, she made another vital discovery when she found the source of the mold infection in her body: after calling in mold experts, they discovered black mold present in her attic and behind her shower, near her dressing room and bedroom, where she spent many hours every day. She immediately had the mold and the source of damp fixed. "I finally got to the root of the problem—the undercurrent was mold," she said. Between that and naturopathic treatments for her liver, her health began to transform.

"Within one week, I was lifting weights again!" she said—something she has not been able to do for a year. "And, I've been able to exercise several days a week and NOT crash."

During a vacation later that month, Patty was able to hike, paddleboard, and Jet Ski, keep up with especially vigorous Pilates workouts and take her dog out for long walks. Her sleep, previously so fitful, also improved.

By the end of October, Patty continued to improve to the point where a biofeedback scan showed her cellular age was now thirty-five—"not bad for a fifty-five-year-old that just went through the perfect storm of heavy metal overload, Epstein-Barr, and black mold."

As autumn carried on, Patty found the energy to start working again and taking on some new clients, and by the seventh month, Stephen, of his own volition, had taken out a subscription to *Townsend Letter for Doctors*—a magazine devoted to alternative medicine. Soon afterward, they had the first positive conversation about his health they'd ever had. "He felt supported and respected and then made an appointment with a new integrative MD to address his health concerns," she wrote. "This is the guy who has said, 'I don't believe all of this stuff!' "

Other areas of her life were transforming too. Patty learned how to create boundaries with her demanding stepmother and how to back off from her stepchildren and avoid allowing herself to be the scapegoat.

In November, the Triton group began sending intention for Patty's family to get along better, while Patty herself began sending intention to "change the script" about her relationship with two of her stepdaughters. As the month wore on, she noticed that she and her youngest stepdaughter, Jessica, were having close phone chats—far closer than they've been for a decade. Then in December, the entire family spent a week together to take a formal family photograph and had the best time they'd had together for years. While in Santa Fe, she and Stephen happened to drive by a house for sale, about an hour's drive from three of the four children in Albuquerque. They'd talked for years about having an investment property but hadn't seriously pursued the idea. The property had six bedrooms and three separate living areas—as though custom-designed for her extended family to visit.

"Since three of us are already in Albuquerque, you might as well retire out here," Jessica remarked as they went from room to room, a sentiment echoed by two of the others. Stephen and Patty ultimately didn't buy that house, but they did find another similarly suitable house nearby. "It's like my group helped me to manifest the perfect house for our family," Patty wrote.

So, were all these breakthroughs down to the Triton group's intention? What most worked for Patty was having to make a public commitment "to the universe," through her group, which forced her to keep looking even more intently for the source of her health issues and work harder on dealing with the cause.

"What I have noticed is that the intentions continue to fuel my efforts for healing," she said. "In other words—I've given my word/put my word/intention out there and it fuels my follow-through. I don't know if it's the intention work or that the intentions continue to keep me very focused and committed to my healing plan," said Patty. They also helped her to forge a new purpose. "I'd been dedicated for years, yet I feel supercharged in my commitment to get on with making a huge contribution to this world. I need to be healthy to do that."

For Patty, the group intention primarily gave her the impetus to uncover the source of her health issues—as they did for Mitchell Dean, another participant, who had suffered from depression for as long as he could remember. As a forty-four-year-old clinical psychologist, he'd put it down to a traumatic birth at Johns Hopkins, where he was born via cesarean section and fed sugar water for three days while his mother recovered from surgery. "This must have set up a very strong freeze response," said Mitchell. "No matter how much I yelled, I couldn't get help or food, which causes a person to start to shut down."

Mitchell was well loved by his family and had been a diligent student who got good grades, but all through his childhood and into adulthood he found it difficult just to be in the world. At times, he would descend into major depression, where he harbored thoughts of suicide all day long. He never acted on his thoughts—he couldn't bear to hurt his parents or, as an adult, his wife and nine-year-old son—but if he saw a bus passing by, he often willed it to run over the curb and hit him. To be a psychologist suffering from depression was doubly difficult, and as an integrative therapist, over the years he had tried everything, from diet and supplements and Chinese herbs to chiropractic, but nothing had seemed to help.

Not long after Helios, his masterclass group, sent an intention to help him with his condition, he was inspired to work with a chiropractor who ran forty-six tests on him. When he got the results back, forty-five were fine, but the forty-sixth showed that one of his liver filtration systems wasn't working. That meant that some of the toxins his body took in were going directly to his brain. Mitchell started a new regimen of Chinese medicine, diet, and supplements, and this time, they worked.

Finally he'd had a true breakthrough; although the depression returned for a day or two now and then, it had subsided. "Holy cow," he thought at one point, "I feel a lot better." But the most profound effect on himself occurred whenever he held an intention for someone else. "It just feels like there is more good fortune that comes my way," he said. "Something in me feels more central, more grounded, more hooked up—like a conduit to spirit."

Alison Maving, a fifty-four-year-old living in Belgium, suffered from vitiligo since 1991. At the time she joined the group, Alison's skin had white patches all over her arms and body. An alternative treatment she'd tried over the years had improved the vitiligo, but nothing had really worked, and although Alison was resigned to the condition, she wanted to find out why it had come about in the first place. During the first seven weeks of the course, while holding an intention to understand the cause of her vitiligo, she came across some articles and books claiming that certain vitamin and mineral deficiencies that she had may be a factor in autoimmune conditions like hers—an association she'd never heard of before.

By months four and five of the course, after she asked the group to focus on healing her skin condition, Alison discovered articles maintaining that vitamin D deficiencies are a cause of vitiligo. She went for a blood test, and although her levels were adequate, as far as her doctor was concerned, they were far lower than those recommended by the articles she'd read. Alison started taking vitamin D supplements. Almost immediately the skin of her arms and legs began to repigment—which she demonstrated by sharing monthly before-and-after pictures with me and her group. By October, her skin continued to regain pigment, and her sister, who also suffered from vitiligo, started taking vitamin D, and her skin began repigmenting too. "This has been a twenty-year search for a cure," she wrote.

A stay-at-home mom, she also had a small craft business, which she was losing interest in. Alison was keen to find her life's purpose and hoped that it might include something related to alternative medicine, a passion of hers. In November, after taking a course in Reconnective Healing in Brussels and other energy healing training, Alison made the decision to become a professional healer. "I feel that I have found my life's purpose," she noted in one progress report. During the same time, all family relationships—another intention of hers—vastly improved. "My relationship with my sister is the best it's ever been," she wrote.

The Morpheus group decided to focus on someone outside their group by attempting to help Laura, a friend of one of their members, regain her health from severe sciatica and breathing problems, which left her with no energy and chronic insomnia. By the end of September, Laura's health had completely turned around—so much so that her pharmacist made changes in her drug regime. Her pain greatly diminished and she began getting up to five hours of uninterrupted sleep. By December, she was able to do grocery shopping, run errands, and cook—all things she had been unable to do. Her health and energy were so improved that she traveled to a national park where she was able to take short hikes.

For Alison, Mitchell, and Patty, intention proved to be an impetus to help them get to the root of their health challenges and begin to overcome them; for Joanne Brockway, the support of the group presented her with the ultimate exercise in trust. Joanne's twenty-two-year-old daughter Jessye, a para athlete, had been chosen by Team Canada to compete in the discus, shot put, and javelin competitions at the 2015 International Wheelchair and Amputee Sports Federation's Junior World Championships in Stadskanaal, Netherlands, in July of that year. Born with dislocated hips, which left her with limited mobility in her legs, Jessye had been introduced to the sports just two years before at a "have a go" day, and had gone on to win two gold medals, break the Canadian women's record in discus throw in national competitions, and become the 2014 Junior World Champion in para discus and shot put.

Joanne had planned to accompany Jessye to the Netherlands and for them to travel around Europe after the event.

During one of our teleseminars after the attendees had been broken into groups, Joanne connected with Iris and Lynette. Although they weren't in the same formal intention group, they'd become intention buddies via phone and email, and before her trip, Joanne asked the two women to hold an intention for their safety and well-being throughout the mother-daughter trip, for transportation to be easily found in "lovely surprising ways," and for their journey be filled with "abundant, synchronous" moments. Iris and Lynette decided to start holding intention for Joanne to receive the finances to pay for the ticket, "and suddenly, unexpected funds began showing up; I kept getting money in amazing ways," she said. Joanne was so impressed that she decided to trust in the power of her friends' intention and made no plans for transportation other than the transatlantic flight.

The first bit of luck occurred when Athletics Canada extended Jessye's return flight date at no extra charge. Jessye's transportation from Amsterdam Airport Schiphol to the sporting event in Stadskanaal had been paid for; Joanne's had not been. If necessary, she could rent a car, costing three hundred dollars for the three-hour drive, but she wanted to avoid spending money if she could and decided not to book anything. "Since Iris and Lynette were holding an intention for us, I was comfortable winging it, trusting that something interesting would happen," she said.

Although from another part of Canada, the mother of the only other female athlete and the only other Canadian parent on the trip happened to be on their same flight to Europe and had booked a car. They got to chatting, Joanne offered to be her traveling companion and split the costs, and the two immediately struck up a friendship. During the games, they shared rides to the daily sporting events and meals when their daughters were busy and, after training, treated both girls for dinners and sightseeing.

Although Joanne had paid for her hotel room entirely through points she had accumulated with Best Western, the hotel manager nevertheless introduced himself, checked her in personally, and gave her a large deluxe room with a giant balcony and a panoramic view of the city—one of the most luxurious double rooms at the hotel.

During the trip, Lynette and Iris kept holding the intention for them both, and even Jessye seemed to benefit. She won a gold medal in discus throwing and a bronze in javelin throwing, but also became the most photographed athlete at the event.

After the competition and some days spent sightseeing in Amsterdam with Jessye, as a surprise, Joanne suggested that they catch the high-speed Thalys train to Paris, a city Jessye had always longed to visit. Even though they purchased their tickets at the last minute on an overbooked train, they ended up with amazing seats in a special glassed-in car, with comfortable chairs and tables all to themselves. After arriving at their hotel, again paid for with points, they were given a lovely large room with a panoramic view of the Seine.

One day, when attempting to visit the Eiffel Tower, they were dismayed to see that a giant line had formed waiting to take the elevator up to the top. Suddenly a security guard called out to them, leading them directly to a back elevator, which lifted them directly to the summit. That luck carried on during boat tours, meals at restaurants, and trips to the various sites; they never had to wait in line for anything, even though it was the middle of the city's heaviest tourist season. When they got lost one night, a taxi suddenly appeared to take them back to their hotel, where on their final morning, the manager gifted them with a complimentary meal.

"It may seem like ordinary experiences, but they weren't," said Joanne. "Many synchronous moments occurred, giving us what we needed, exactly when we needed it, as we enjoyed the totally positive connection with people we met. We felt safe, even when we were lost at night." And best of all, the entire journey ended up costing them no more money than the cost of two round-trip train fares, daily subsistence, and Joanne's flight.

For Karen Hayhurst, a forty-nine-year-old single parent with two daughters, the intention of the group acted as a springboard, giving her the courage to cut back on a soulless job as a driving instructor, which was only paying the bills, and return her to the work she really loved. "I always felt the energy work was my purpose, but the driving instruction was my job," she wrote.

When she started the yearlong course, Karen was suffering from a painful lower back and sore knees from spending long hours in a car with little time for anything other than eating and sleeping. The lack of exercise was also causing her to put on weight, even though she wasn't eating much during her long working days. Although she has many friends, her overfilled work and home schedule left her with little time to socialize.

By the end of the first seven weeks of the course, the pain in her back and stiffness in her neck had gotten so bad that she had to leave her driving instructor job.

During one of our teleseminar calls, after being placed in an intention circle and in the midst of sending intention to one of its members, Karen was overwhelmed by the loving vibes of the group, coming to her in waves. Just after the group finished, she looked down and there was a paper in front of her about energy research. The words just popped into her mind: *How could you leave it?* "The tears fell and I knew it was time to get back to it," she wrote.

Since Karen could do very little driving due to the pain in her neck, she had the space and time to return to her energy research and developed an online course during her recovery. "I learned a ton about the actual producing, quality of video and audio plus how to edit videos," she wrote. "In addition, I started collecting tons of research on energy work. I felt exhilarated."

During Karen's time off, she finally had time to have coffee with friends and make new ones; she even took a day road trip with a good friend, which she hadn't allowed herself to do for years. For the first time, she was able to spend uninterrupted quality time with her girls and to resume her daily call to her mother, which had been impossible with her long work schedule. By the summer, Karen's neck injury was healing and she was able to have a daily morning walk, which helped her to lose weight.

The downside was that she had no paycheck coming in and had to rely on her savings. However, after finding out about Karen's injury, her estranged father, to whom she hadn't spoken for seven years, made contact and sent money to help her through this patch. Karen wasn't just grateful for his help; the renewed contact "smoothed the waters between us."

When autumn arrived, besides enjoying increasing closeness with her mother, daughters, and friends, Karen started hearing from past contacts, who began passing along research links. Various growth opportunities for her started springing up. "I find myself now as the research hub for many of my energy/holistic colleagues," she wrote. "I have a website up and running, a regular blog, and growing subscriber list." Eventually she did go back to doing some driving-instructor work to pay the bills but did not allow it to take her over, as she had in the past. She went on to earn a bachelor's degree in holistic health sciences and is currently working on a master's and PhD in natural medicine.

Like Karen, Melissa Fundanish, a fifty-year-old from Tega Cay, South Carolina, joined the masterclass with the specific intention to find a new career opportunity. At her job at the start of the masterclass, she had an unhealthy relationship with her manager and felt caught in the middle of two teams who seemed unable to work out how to collaborate for the common good of the customer. Furthermore, the product she was supporting had a limited life and was likely to be obsolete in the next few years. "I am having trouble determining the type of position I want and how to go about looking for it. I feel stuck," she wrote on her progress report in early summer.

In July, after setting an intention with the group to find meaningful work, Melissa received an email from a colleague. Melissa knew the woman was experienced in conducting job searches and plucked up the courage to ask if they could have a private chat. When they met, Melissa confided in her about her desire to move jobs. As it happened, the woman was looking to fill a particular job at that time that would have suited Melissa perfectly. "I asked for a job description, thought about it overnight, and decided to apply," Melissa wrote in her next monthly report. "She fast-tracked me through the interview process with four interviews and a presentation within a week. I received an offer the following week for much more money than I expected."

Melissa began her new job in August. "My new job is amazingly wonderful," she wrote. "Two months in, and I am loving my manager, my peers, my employees, the culture and my responsibilities. I really didn't think it would be possible to find something I would enjoy and feel challenged by."

Besides her job, the Proteus group focused on a very specific request: having Melissa sell her BMW M3 for five thousand dollars and specifically to someone who loves BMWs. "Out of the blue, I received a call from a gentleman in Colorado. We had a nice conversation and he said to me, 'Okay, I want to buy the car!' And I said, 'Great, what do you want to pay for it?' And he said five thousand dollars. I was also very happy that it was sold to a BMW M3 enthusiast."

Melissa decided to set a similar and highly specific intention for her sister's car, which had been on sale for three months with no takers. Three weeks after Melissa's intention, her sister sold the car for ten thousand dollars, as she'd intended, and to another BMW aficionado.

For Melissa, the group helped her fall back in love with her life. "I am finding that overall I feel like I am in a wonderful flow with life. Things happen more easily and I am enjoying the flow." She and her sister enjoy a closer and deeper relationship, and she is meditating regularly. She even sent an intention that she find someone to begin a relationship with. "Within a week or two, a friend of my sister's scheduled a lunch for the two of them along with myself and his roommate. His roommate and I were surprisingly compatible. We had a first date and really enjoyed ourselves."

Robert Morales, age sixty-seven, from Beaumont, California, also enjoyed improved health as a result of working within his Helios group. At the start of the course, he had issues with his heart, prostate, pancreas, left knee, urinary tract, and sleep. He'd also been diagnosed with type 2 diabetes. He asked the group to send intentions for his health to improve. In October, Robert also asked for assistance from some of the group members to resolve the pain in his left knee. Within approximately eight to ten days, the pain in his left knee subsided. "To this day I have no problem with this knee," he wrote.

Then, on December 9, he came down with flu—partially, he thought, from tiredness and working long hours. Because of the flu, he couldn't attend his usual Thursday sessions with Helios and requested that they send him intention to get better.

On December 11, he felt so much better when he woke up that he headed off to work. It was as if a huge weight was lifted from his shoulders. He was able to sleep better; his heart was no longer palpitating and arrhythmic as it usually was at night. His prostate symptoms were notably absent—so much so that his sleep was interrupted only once that night and the nights that followed, and for the first time in a long time he could eat meat without any residual reaction. In fact, for the first time in many years he could eat anything without a reaction. "I was feeling absolutely normal as if I had no problems whatsoever. I felt healed and energetic. This was one of the best five days I have had in a very long time."

Robert and his wife were experiencing financial difficulties in paying their bills, so he asked some members for intentions to send assistance. Within less than thirty days, he received a check from their bank for $2,475. "We knew that we would be getting something back from our bank as a result of buying a home this year, but we were unaware of the amount and date."

By the time Beverley Sky Fulker asked for the group's support to improve her finances, she was down to her last £200. She had a chance meeting with a person who informed her that anyone who had previously worked for Lloyd's of London insurance could apply for assistance if they needed help. Beverley had worked for Lloyd's, so she applied. Although they are highly selective about applicants, they chose Bev, "and they sent me a rather beautiful check," she said.

Born with a port-wine stain on her face, Bev had dealt with bullying as a child and for years had plastered her face with camouflage makeup and considered plastic surgery. As an adult, she'd decided to set up a website to offer inspirational stories and advice to encourage others with scars or birthmarks to feel confident and positive about themselves. The money from Lloyd's also came in handy in helping her to update her site (LoveYourMark.com). "Just in time," she wrote. "Just when I needed it."

Mitchell Dean found that with the group's help, he healed not only his lifelong depression but also "exceptionally painful" issues contributing to it: "I've made more progress in the last year than in the forty-four years prior, by a lot," he said in an interview. "So much so that I've finally moved into healing some other issues as the primary health goal, and those are moving already too." He overcame the writer's block preventing him from writing a book he'd been planning to write for years, lost fifteen pounds, returning to the weight he'd been in high school, and got in better shape than he had been for years. Mitchell also got back into singing and playing guitar, which he hadn't been doing regularly for decades. A mindfulness technique he had taught his patients for many years suddenly got "discovered" by a well-known actress, who offered to help Mitchell get the word out.

Andy's group intention helped her to consciously "uncouple" from her husband with a minimum of pain and disagreement. They'd agreed to divorce but taken no concrete steps to end their marriage. Within the first few months of her intention group, she emailed a collaborative divorce attorney, and their first meeting went well. After a meeting with a second collaborative attorney, they decided to hire their own lawyers, at which point Andy learned that her husband has a serious girlfriend. This proved to be an important impetus, as they then agreed to tell the children they were divorcing and Andy's husband moved out. With the support of her group, she was able to stay strong despite insensitive comments from other family members.

"Our communication is deeper and stronger and more open than it was in any of the years of our marriage," said Andy. They worked on a "collaborative divorce," agreeing not to litigate, and their attorneys were so amazed by their ability to hammer out the details of the separation in everyone's best interests that they asked to share their system with their other clients as a model for what was possible in divorce.

Margaret, a probation officer in California, where drug use remains a serious problem among former offenders, asked the group to send intention for a 50 percent reduction in the positive urinalysis results for her clients undergoing random drug testing—and promptly got the result she was looking for.

Trudy regained some of her hearing.

After the Achelous group's intention, Amanda's daughter-in-law, who had had two prior births with days of difficult labor, had her third child fifteen minutes after arriving at the hospital.

Rose White sold her house in two weeks and found her dream home.

Month by month the list of extraordinary transformations carried on. I had one last experiment to do, one final peek inside the clock's inner chamber, to find the mechanism at the heart of it all.

Chapter 21

The Power of Eight Study

Dr. Guy Riekeman, president of Life University, likes stirring things up, a fact he immediately makes clear on the university website's home page by pronouncing this as a place for "visionaries relentlessly committed to disruptive social innovation." As he and his fellow faculty members see it, they are setting off a revolution in health care, with their students leading the charge away from the "sickness care" of the conventional medical model to one of holistic wellness. Riekeman and his other chiropractors at Life U are from the camp of chiropractic who believe in vitalism, that all systems of the universe are, as the university puts it, "conscious, self-developing, self-maintaining and self-healing." The vitalists view their job as simply removing the impediments, in the form of badly positioned spinal vertebrae, blocking the free flow of this energy, like so many fallen branches across a railway line.

A craggy-faced seventy-year-old, Riekeman is a giant in the chiropractic profession who, after taking over the presidency in 2004, grabbed the university by the throat and within a decade transformed a small collection of concrete buildings in a wooded backwater of

Marietta, Georgia, into the largest chiropractic college in the world. I sat with Riekeman and a number of other members of the faculty one evening in April 2015, after presenting at Life University, and over a particularly memorable bottle of red wine from his private collection, Guy volunteered the university's services to study what was going on in my Power of Eight groups, placing the departments of biology and psychology, with all their scientific measuring equipment, at my disposal.

I was overwhelmed by this generosity. This was exactly what I'd been looking for since 2007: a respected university willing to do an experiment on my Power of Eight circles.

To my mind, the most important subject of our study was not discovering whether we could affect a receiver, but examining what was going on inside the senders. Guy put me in touch with Dr Stephanie Sullivan, a neuroscientist and director of the Dr. Sid E. Williams Center for Chiropractic Research, who has a great deal of experience carrying out scientific research. With her assistance, we settled on a simple study of individuals participating in a Power of Eight group made up of volunteers from the student body. I would offer successive Power of Eight groups some simple instruction via Skype or a YouTube video, one of the group members would volunteer as an intention target, and the rest of the group would send intention to him or her, just as I'd done in my workshops. Stephanie and her team would examine the brain patterns of one of the intenders via a qEEG, or quantitative electroencephalogram, the standard equipment for measuring different brain wave patterns, before, during, and after the group intention. All the participants would fill out forms assessing their moods before and afterward so we could assess any changes.

In order for our study to have scientific credibility, Stephanie planned to conduct the trial seven times with a different Power of Eight group made up of different individuals, each of those targeted for brain wave measurements first-timers in this kind of intention.

She'd then compare the results of our qEEG readings with those that had been recorded in individuals participating in studies like Andrew Newberg's and Richard Davidson's to see if my intention procedure produced any distinct difference. While it wasn't a fully controlled trial, we would still have a decent preliminary study that might give us certain clues about why participants in my experiments had recorded such life-transforming effects. Later on, we would examine immune system markers and other biological activity in both senders and receivers, to find out if there were any other substantial changes going on.

Stephanie sent through the first results in early February 2016. "So far, the results are quite amazing," she wrote. Our participants showed evidence of immediate and major global brain changes that were considerably different from normal, she said.

A few months later, after carrying out and analyzing six of the seven studies (one proved unusable), the Life University research team discovered a scientifically significant decrease in activity in the right temporal lobes, the frontal lobes, and the right parietal lobes of our participants during the intention sessions, an almost global quieting of the brain occurring in several frequency (or brain wave) bands. In fact, our results were the opposite of what occurs in general meditation, which tends to cause an increase in alpha and theta brain wave power for the majority of the cortex; alpha waves *decreased* in our participants. The greatest changes occurred across the entire right parietal lobe of the brain—the part that distinguishes our sense of self from everything that is not self—the temporal lobes, including occipital region, which usually relates to vision—and the frontal regions of the brain, relating to executive processes like planning and decision-making. The temporal lobes are also associated with memory, visual representation, and auditory processing. The fact that our results were statistically significant and consistent across the participants' studies would suggest that they weren't due to chance, said Stephanie, particularly as they occurred immediately, during just ten minutes of sending healing intentions, and among people who'd never engaged in a Power of Eight intention circle before.

Chasing enlightenment by engaging in contemplative exercises ultimately is a self-oriented activity, reflected in an increase of activity in the "self" aspects of the brain. "When a person chooses to seek Enlightenment through a specific practice—be it Eastern or Western, religious or secular—activity initially begins to increase in the frontal lobe when she begins to meditate or immerse herself in contemplative reflection," wrote Newberg. "We also see in our brain-scan studies an initial increase in activity in the parietal lobes. Our awareness of our self in relation to the world or object of meditation is increasing, and parietal activity helps us to identify our goal and move toward it."

But in our Power of Eight groups and Intention Experiments, moving away from the self and focusing on the "other" immediately reduced activity on many of the areas related to self, particularly those on the right side of the brain, which, aside from creativity, is associated with negative thinking, fear, worry, and depression. Specifically, we found diminished activity in the right prefrontal cortex, which could indicate a shift away from higher stress and anxious states and an improvement in emotion, wrote Stephanie. Indeed, on a Brief Mood Introspection Scale, a standard psychology rating test she gave to the participants of each group before and after the Power of Eight intention circles, significant improvement was noted both for the overall mood score of the participants and also in calm and relaxation. Stephanie gave the participants a standardized scientific test measuring any changes in pain. Although we didn't observe any significant trends, and our Power of Eight groups were a uniformly healthy group of young students, a number of the members reported pains of various kinds—migraine, joint pain, bad back—and their pain spontaneously disappeared.

The brain waves of our participants resembled the very signatures of many of the groups Andrew Newberg studied who attempt enlightenment, but mainly through a process of surrender—the nuns and monks, mediums, Sufi masters engaged in chanting, and even, to some extent, the Pentecostal Church members speaking in tongues. In instances where the attempt at enlightenment is not a self-oriented activity, as it wasn't in those instances, and in the case of our Power of Eight groups, the frontal-lobe activity tends to immediately fall away, as the person begins to merge with the object of contemplation. Our Power of Eight study also showed evidence of an increase in coherence between the parietal and frontal lobes. When applied to the brain, "coherence" means the degree of communication between the different parts of the brain. In our case, although activity was diminished, the brains of our participants appeared to be operating as a greater whole. Participants also demonstrated reduced activity in the sensorimotor strip and association areas, the location for sensory and motor processing, the area of the brain that gives meaning to sensations, including music. This would suggest that the Power of Eight members had entered another dimension in which they were far less aware of their immediate surroundings.

This offered more evidence that these changes didn't have much to do with the Reiki chant music, since the participants experienced a decrease in all the parts of the brain that recognize and process music. And the decrease in the occipital region—related to vision—may relate to the fact that their visual attention was directed internally to a visualization of the person to be healed.

Our group members appeared to be experiencing an altered state of consciousness, just as Newberg's nuns had done. But the Power of Eight participants weren't in Holy Communion with God; they were in Holy Communion with each other and the person or situation they were trying to heal. The Life U study suggested that the participants in our global experiments and Power of Eight groups were experiencing something akin to a moment of ecstasy, which then may have proved transformational in their lives. But unlike Newberg's nuns, monks, or Sufis, the process hadn't required priming—an hour of intense chanting or reflection to achieve that state—or years of devoted practice. With those subjects, and indeed in most instances of contemplative prayer, says Newberg, "typically it took about fifty to sixty minutes for them to create these same kind of neurological changes."

Something vastly different had happened to my participants. They'd entered into this state within a few minutes of beginning a Power of Eight group or an Intention Experiment, and their experience of enlightenment was not only immediate but both unexpected and uninvited. And unlike your typical religious or indigenous experiences, there had been no mantras, no fasting, no self-denial or deprivation, no sweat lodge, no yoga or prostrations, no speaking in tongues, no icons, no ayahuasca, no "great effort of the mind," as Saint Augustine had once described. In fact, there'd been no real effort at all; the experience had been mostly out of their control. They didn't turn it on—their involvement in the group intention just made it happen. The only initial inducement was the Powering Up—the short mindful meditation ritual we use in all our Intention Experiments. Every person examined in our study was a complete novice who'd never practiced Powering Up before; the most experience they'd had was intermittent meditation, and their only instruction manual had been a thirteen-minute YouTube video I'd made describing how to proceed. In our Intention Experiments and my workshops, the vast majority of our participants also had never practiced my Powering Up before. Although those who took part were mostly experienced meditators, as they recorded in their surveys, for most of them this experience was qualitatively different from ordinary meditation. In every case our participants had been transported into that state in an instant.

There was no other conclusion I could draw. Sending altruistic thoughts of healing in a group was a fast track to the miraculous.

After studying the EEGs of a batch of pairs of guitarists, the psychologist Ulman Lindenberger and his colleagues at the Max Planck Institute for Human Development in Berlin, Germany, discovered that when two or more people play music "with one accord," their brains begin to mimic each other. The brain waves of each pair become highly synchronized and "in phase"—that is, their brain waves begin peaking and troughing at certain key moments. Entire areas of the two brains create synchronized patterns, particularly the frontal and central regions, but also the temporal and parietal regions, those parts of our brains that govern our sense of self in space, and in this instance, the synchrony suggests the guitarists begin to feel a sense of unity with their fellow guitarists.

The same team went on to study guitarists who were improvising together and discovered what's been called a "hyperbrain pattern"—the tendency of the brains to work in tandem so closely that they come to resemble a single giant brain—particularly when both guitarists are playing at the same time. Other scientists at the University of Lancaster in the United Kingdom and the University "G. d'Annunzio" of Chieti-Pescara in Chieti, Italy, have discovered the same results when studying shared thinking—or what they refer to as Team Mental Models—between groups of jugglers. The juggling pairs develop not only a hyperbrain pattern, but also coordinated heart and breathing rates.

In the case of my Power of Eight groups, this was no longer a collection of separate individuals. The borderline separating them had been erased. This was a supercharged hive, a supergroup. They weren't just connecting—they were merging.

But the changes experienced by the Power of Eight groups could also be radiating out to their environment. Konstantin Korotkov had perfected a sensitive device he'd playfully christened "Sputnik," after the first Soviet satellite space launch in 1957. His device was a bit like Roger Nelson's entire Global Consciousness Project configuration rolled up into a single machine, as Konstantin claimed that it was capable of measuring environmental influences on human emotion.

Sputnik had been developed as a specially designed antenna for his GDVs, which Konstantin liked to refer to as an "integral environment analyzer." Coupled with the information delivered by his GDV, the purpose of this highly sensitive device was to measure any changes in the atmosphere relative to any changes in the people occupying that space. Konstantin claimed the little sensor could pick up the capacitance, or ability to store charge, of the environment through its extreme sensitivity to changes in environmental electromagnetic fields.

As human emotions are related to the activity of the parasympathetic nervous system, any changes in that system also change blood circulation, perspiration, and other functions, which consequently change the overall electrical conductivity of the body. Aware of the body of evidence demonstrating the effect of solar activity, tectonic disturbances, and tensions, and the ambient electromagnetic field on human health, Konstantin maintained that the reverse was also true: when a person experiences a change of emotion, it will affect the electricity of the environment, which in turn will be picked up by his Sputnik sensor.

"Changes in the functional state of the human body leads to a change in . . . the field distribution around the body, the chemical composition of the ambient air due to exhaled air and emissions of endocrine substances through the skin," he wrote in a paper about his invention. It was his theory that his Sputnik was capable of picking up even the most subtle of these environmental charges.

Konstantin had spent a number of years testing the device during expeditions to Peru, Colombia, Ecuador, India, Myanmar, Siberia, and elsewhere before becoming satisfied that the device was sufficiently sensitive to assess local environmental conditions and their idiosyncrasies after discovering sensitive sensor signal variations during sunrise and sunset or prior to a thunderstorm. In 2008, he'd taken a series of measurements with it in a variety of spots in Russia—Novosibirsk, Berdsk, Irkutsk, and Abakan—using seven independent Sputnik devices during a total solar eclipse. All seven devices showed similar curves of activity before the eclipse, with all stabilizing similarly after the event was over.

His most intriguing claimed effect was the ability of the device to measure the subliminal psychological and emotional reactions of groups of people. He'd tested this during a variety of group gatherings—religious ceremonies, yoga exercises, group meditation, musical performances, and even public lectures—and discovered statistically significant changes in the device that correlated with the duration of the events and the group's collective emotion; the higher the changes in his Sputnik signal, the greater the emotional charge of the room. In one study, like Roger Nelson and his REGs, he discovered major changes in the machine's output during periods of intense meditation. He'd also gone on to demonstrate the effect of subliminal emotion on a room's charge with one simple study of the impact of low-intensity sound on a group of student volunteers. They'd been asked to come into a classroom and simply work on computers, while unbeknownst to them, Konstantin turned on a device emitting a low-intensity 20 Hz sound—on the border of the human range of hearing but enough to be subliminally disturbing.

After the study was finished, a questionnaire assessing the students' mental and emotional state, including their perception of their health and mood, unquestionably showed that they'd been stressed during the experiment, and their changes mirrored the changes registered by the Sputnik. The same changes did not occur with a control group of students under the same conditions but without the sound played or even by a third Sputnik, exposed to the same 20 Hz sound but placed in an empty room.

In March 2017, I ran another water experiment with Konstantin in which we asked the audience of one thousand in Miami, Florida, to send intention to a bottle of water connected to a computer installation in his laboratory in St. Petersburg. Even though we were more than five thousand miles away, the measurements clearly showed a significant effect on Sputnik, a major lowering of charge surrounding the water.

During two workshops of mine, when Konstantin was present, he turned on Sputnik and also measured some of the participants before and after our Power of Eight groups. In both cases, stress levels had dropped considerably in the individuals, and a change of charge in the room had clearly been registered. The effects of the Power of Eight groups were affecting the group members but also radiating out, sending out waves of good will.

The same had happened when Roger Nelson turned on his REG machine during our Power of Eight groups in Italy. Roger had been at two conferences with me in Bologna and then in Rome, and in both instances, when I was running the Power of Eight circles, I'd asked him to turn on a REG machine he had in his computer. Each time, the effects grew greater and greater as the groups carried on, some freakishly pronounced movement away from randomness to order.

Sending and receiving, receiving and sending.

My masterclass groups continued to meet together weekly, even after the year was over, miracle vortices of transformation. Teri had a past real-estate client call her out of the blue, saving her financial situation—"from just a step shy of homeless to having a steady stream of income as a real estate broker"; Linda's Grow Food Earn Money tour got the promotion the Triton group intended, and a major college chose to teach her methods; out of the blue, Melissa's estranged father sent her a check for ten thousand dollars, and she received another ten-thousand-dollar windfall when a past company she'd worked for briefly bought out her pension; Yoly's relationship with her husband underwent a major transformation, and he became more supportive of her wish to pursue her entrepreneurial interests rather than simply focusing on her role as a wife and mother; Laureen had invested in DynaCERT, a company involved in reducing diesel and large-engine emissions, and after her group's intention, the company realized a 558 percent increase in share price and 677 percent increase in market capitalization, and got ranked as the top-performing company in five industries. Besides the major windfalls, there were smaller successes too, and the miracles started showing up with greater frequency after I urged them to stop intending for themselves. Julie established a regular meditative practice, "for the first time in my life"; Nancy began to lose the twelve pounds she'd wanted to shed; Andrea made it through the Christmas holiday without any quarrels with her mother; Judy's food movement advocacy group got the help she needed; Kristi's digestive issues disappeared; Marie began attracting new tax clients with no effort; Bev reconciled with her estranged brother; Iris's chronic congestion began clearing up; Martha's insomnia completely resolved. Family members, friends, even pets also benefited from the intentions of the groups; Barbara's husband began working on a new project he hadn't attempted for years; Laureen's husband's condo got sold for the asking price; Elaine's sister-in-law, who suffered liver failure, ended up not needing a liver transplant, against all expectations, and is recovering and her brother-in-law also managed to avoid planned surgery for a tumor on his esophagus; Karen's mother's diabetes is far more controlled and she's eating regularly for the first time; Melissa's kitten, born with underdeveloped lungs, is now close to normal; Jane's horse Calypso was saved multiple times from being put down. Besides having business opportunities show up in unique ways, Marnie made a "huge shift—a feeling that is hard to explain, that really all is well. Satisfied in my life and its course. Real joy and gratefulness."

Of the 150 regular attenders of the groups, nearly every single one had made some sort of major shift. Many, for the first time, found their life's purpose or improved their relationships or discovered how they were self-sabotaging. "I take responsibility for my part in interactions and try to remember that I come equipped with a pause button," wrote Joan Johnson.

My masterclass groups had conducted their own experiments, as I threw out more and more challenges to them. The Proteus group set up their own experiment to increase rainfall in Charlotte and the surrounding area in order to nourish the foliage, with Melissa investigating the monthly rainfall for the year and the deficit from the monthly average. When they started, Charlotte and the surrounding area were suffering from a heavy drought, a yearly deficit of 14.34 inches of rain; after beginning their intention in September, a slow steady rain began to fall. By the second month the area had surpassed the average monthly rainfall, and by December, in three and a half months, the area enjoyed more than 12 inches of rain, largely making up the year's deficit.

Unbeknownst to Marie's brother-in-law, who arrived for Christmas with a bad leg ulcer and in excruciating pain, Morpheus decided to put their healing intention into water, which Marie sprinkled into his drinking water. After consuming it, he didn't have another day of serious pain during the holiday period.

What is it about groups that helps to manifest all these changes so that close to 100 percent of those 150 experienced personal miracles all in the same year? What are the odds against the chance of all this being mere coincidence, the inevitable changes in circumstances that occur over time?

I no longer ask that question. I'm content simply to be a kind of messenger, a reluctant apostle of the mysterious alchemy of groups.

Most of our attendees talk about the invaluable support the group gave them to initiate change in every aspect of their lives. Many cry when they talk about how much the group means to them—"my intention family," as Ellen Bernfeld referred to her group. The Power of Eight group is there, said Ellen to help her "keep getting back on that horse every time I fall off."

"Practicing the intentions always takes me to a place of receptivity in myself where my 'yes, buts' become quiet," said Lissa Wheeler. "My mind becomes quiet and the 'trust' is more of a knowing on a visceral level that the intention is unfolding. The group connection carries me into this space. Sometimes I would be resistant to settling myself down enough to focus and would sit there almost pouting like a grumpy child, but within five minutes of being willing to stay with the group focus, my brain literally felt like I'd taken a medication that took me into an altered state. It was as if we became one brain together. The group focus felt like a soup that slowly changed my brain."

"Something about the process of holding intention for others, being randomly connected to a batch of strangers, whom you haven't had classes or kids together with, is very profound," adds Mitchell Dean. Like most of the masterclass members, Mitchell has never met his group members, but feels intensely close to them, particularly to Robert Morales, the one member he's kept up with consistently since the masterclass ended. Mitchell and Robert are in touch almost every day to help each other with all sorts of issues. Mitchell would always email, as he did one night when he was having trouble sleeping, and, in that case, as always, Robert wrote back the following day: "Don't worry—at 4:30 a.m. I had you covered." And now when he hears that someone has a problem, Mitchell himself says: "I'll work on that for you."

"It's about the process of giving," says Mitchell. "I feel better all day when I'm working on others. And it's not just gratifying to see someone happier—I'm fortunate to get plenty of that in my work as a psychologist. There's something more there that I don't understand. My own system, my life, just works better when I'm of service in this way."

Elaine Ryan, a sixty-one-year-old from Katonah, New York, perhaps put it best: "One evening I saw a visualization of our group Helios as separate pieces, then the pieces started to merge into this nucleus that was growing and becoming unified and one, Helios being the nucleus. So each week we build on the last and become more unified, stronger and more committed to the focus of our intentions.

"I find myself thinking of members in the group or of the group as a unit and a part of my life, more as a unified energy field as well as individuals with their own paths."

A few years before, in September 2012, I had tried one more major Peace Intention Experiment in the midst of the mudslinging match of that autumn's presidential election. I'd wanted to do a "Heal America Peace Intention Experiment," and had targeted what I'd considered the most violent place in America: the US Congress. I'd first played around with it on *Coast to Coast AM* talk radio show in June and then in a single day's event over Gaiam web television in September, directing our thousands of viewers to send intention to the two counties surrounding the Capitol building that were suffering ever-growing levels of crime, then sat back and waited an entire year to compare the police figures for twenty-four months before and twelve months afterward.

The following September we analyzed the data: violent crime, the target of our intention, had decreased by 33 percent, starting from the September of our intention, bucking the trend from the two years before, while property crime continued to increase. And the day after our *Coast to Coast* intention, for the first time since they'd known each other, Republican Speaker of the House John Boehner had hugged his sworn enemy, Democratic former Speaker Nancy Pelosi.

Did we do that?

Had "we" done any of it?

Short answer: Who cares?

I had more than my usual two sources—the testimonials of thousands of participants, and even a scientific study—to demonstrate that group intention has an extraordinary effect on both the senders and the recipients, but I am no wiser about the "why," the exact cause of these miracles. Is it the intention itself, the amplified power of prayer in a group, or just the fact of making a public declaration of intent, as Patty Rutledge believed?

Something about the promises we make to each other may carry more weight than the promises we make to ourselves. They give us the courage, like Guy's vitalists, to remove the branches lying across our tracks with greater ease. A statement in the presence of a small group is a contract we make with the universe—to do and be better than we presently are. There is also the power of support and connection, a condition as necessary to the human spirit as oxygen is to the human body. The most fundamental promise we make to each other, the most basic of our social contracts, is to support each other through adversity. *I will be your witness.* At every point of our lives we need to know, that somewhere out there, somebody's got our back, and this knowledge becomes a larger certainty in our lives when a group of strangers connect together to heal us.

I still like evidence in my work, but over the course of studying this phenomenon, I've lost my skepticism, my need to tease out some scientific basis for everything that just cannot be rationally explained. Some things in our lives are just beyond our explanation or understanding, and when people come together, miracles just happen, miracles that cannot be reduced to the sum of certain facts and observable data, the workings of the vagus nerve or brain. I've come to believe that miracles aren't individual but the result of collective forces, especially when we move past the puniness of the self. I've given up trying to explain magic. It's enough to show, even in small glimpses, that it's there.

I've witnessed many miracles with this project: evidence that our world and our innate capacity is far greater than that envisioned by

Newton or acknowledged by modern scientists. I've observed first-hand that consciousness is a collective activity, with the ability to traverse time and space, and that minds connect from any distance when focused on a single point. Connection has nothing to do with proximity and everything to do with the collective capacity to create. All it requires is making a statement *homothumadon*—with one passionate, ecstatic voice.

I've seen the extraordinary power of a small group to create hope and healing in the lives of every group member. I've understood that the most powerful transformational state of all is altruism. Moving away from self-help is our most potent healer. I now believe that group intention could indeed heal the world, but not in the way that I first imagined. The target doesn't really matter. What worked for Mitchell Dean and many others was letting go of any attachment to an outcome. The healing is all in the participation, the desire to pray with one voice.

Intention is the more secular version of prayer, different in its specificity and belief in the individual power to manifest. Rather than leaving it up to God ("Thy will be done") we recognize the power in ourselves as creators and attempt to take charge of our own fates. We have been raised to think of prayer as highly individualized: you having a private word with your maker. For me it's now evident that we amplify that private word in every way when we pray as a communal act. As one person requests a healing, so our own need for healing reverberates deeply inside us. We make a public commitment to each other to try harder the following day. Each time we participate in a healing, we also heal one small part of ourselves.

French anthropologist Laurent Denizeau, a teacher at the Catholic University of Lyon who studied healing evenings organized by the International Association of Healing Ministries, once referred to ritual healing groups as "soiree miracles," suggesting that the coming together of people as a group is a necessary factor in miracle healing.

Although the pastor will work up the crowd with his own and others' healings and invoke the Holy Spirit, in Denizeau's view, "it is not the act itself which creates healing, but the fact of it being realized in an assembly of prayer."

> *En ce sens, la maladie est une épreuve de soi mais aussi une épreuve relationnelle, que ces assemblées prennent en charge. Prendre soin du corps, c'est prendre soin du lien qui le construit comme sujet. Le corps malade, espace de rupture dans la définition de soi (marquée par sa relation aux autres), s'inscrit ici dans un corps plus vaste où la maladie n'est plus l'unique espace commun. Cette sociabilité autour de la guérison agit comme issue de secours du sens.*
>
> In that sense, illness is not only a personal trial, but also a test of the relationship of the group itself that those gatherings support. Caring for the body also means caring for the bond that makes it the target of a group. A diseased body, a ruptured space according to the individual's definition of self, becomes part of a much larger body when viewed in relationship to others, where disease is no longer the only common aspect. When a social connection is part of the healing process, it essentially acts as an emergency safety exit for that small definition of self and gives access to a higher meaning.

Laurent Denizeau is saying is that illness is part of that smallness of the self—a distinct and separate entity—but in the presence of a group, the individual recognizes him or herself as part of a greater whole. The sickness is identified essentially as foreign matter in this perfect unity, like a splinter sticking out of a finger, at which point the group, like a giant tweezer, gently helps it to be excised.

One day when I was listening to the song "One" by U2, I was suddenly struck by the simple wisdom of the words "We get to carry each other." The song of course focuses on how we are still "one" even if "we're not the same," but I realize now that the line about carrying each other doesn't refer to obligation; it's about privilege. With the opportunity to carry each other, we are given the opportunity to be healed.

When I think of all of my intention work, I think of what Jesus may have been trying to tell us. No matter whether you are religious or, like me, have a more secular sense of spirituality, his words continue to resonate. Don't play small when it comes to healing yourself or healing the world. This is too big an enterprise to attempt by yourself. Find your truest self and your greatest power in numbers.

PART II

*

Creating Your Own Power of Eight Circle

Chapter 22

Gathering the Eight

Now it's time for you to test out the Power of Eight in your own life. What follows are instructions on how to create a group of eight people who will meet regularly in person or virtually. It is not necessary to be physically present with the members of the group. A virtual connection, in my experience, works just as well. It's also not strictly necessary to have exactly eight people, but eight is the optimum number. I would suggest that your group be no fewer than six and no more than twelve so that you have enough of a critical mass to feel like a group, but not so many that you get lost in it.

Assemble a group of eight like-minded friends who are open to the possibility of healing and intention. You can use a book group, church groups, or the members of your neighborhoods.

1. Ask if any of the members of the group with a healing challenge of some sort (emotional or physical) would like to be the target of the healing intention. Allow the person nominated as the recipient to describe their problem in detail.

2. Spend a few moments talking over and designing the intention statement that you will all hold together.
3. Gather around in a circle. Either join hands or place the nominated subject in the middle of the circle, as every other member of the group places one hand on the subject, like the spokes of a wheel.
4. Begin by having each member of the group close their eyes and concentrate on inhaling and exhaling. Each should clear their mind of any distractions, then hold the intention statement in their mind while imagining, with all five senses, the intention recipient as healthy and well in every way. All members should then send out the intention through their hearts. The intention recipient should remain open to receive. (Follow the techniques of "Powering Up" beginning on page 240, and for full instructions, from *The Intention Experiment*).
5. After ten minutes, gently end the healing intention and have everyone take a few moments to "come back" into the room. First ask the intention recipient to describe how they feel and if they have experienced any changes, positive or negative. All the other members may then take turns sharing experiences. Take note of any feelings of palpable oneness and also any improvement in the condition of both senders and receivers.
6. With time, begin to select targets outside your group.
7. Keep careful note of any monthly progress in your life: your health, your relationships, your career, your life's purpose.

Making the Eight Virtually

Here are some ideas for starting out:

1. *Create structured times and frequencies for meeting and stick to them.*

   Decide as a group whether you want to meet daily or weekly (I'd recommend at least once a week) and keep to a set time every week. Then decide if you want to meet online, as an audio group or as a video group. All three are possible on Google Hangouts or Skype.
2. *Elect one web-savvy person on your team as the go-to person for problems on Skype, Google Hangouts, Zoom, or some other online facility for virtual group meetings.*

   This person can help anyone who is struggling with the technology to get up to speed.
3. *Before the meeting, write down your major intentions for the month or year* and take turns sharing them during the meeting.
4. *Begin the meeting by sharing who you are and what you hope to achieve by your participation in a group* for the rest of the year.
5. *Start off with questions, sharing, and discussion* about aspects of the work you've learned in this book and other books, like *The Intention Experiment* or *The Bond.*

   This would include becoming more conscious of your thought processes and the implications of the fact that we are sending and receiving information at every moment. What does that really mean to you? How does that impact the success or failures you've experienced in your life?

6. *Include time for questions*, which you should address to the group.
7. *Include time for practice sessions.*

   For the first two sessions, break into pairs and practice sending and receiving the mental image of a simple object that holds some special meaning for you—positive or negative. Receiver: try to intuit not only the object but the emotion held by the other person about that object. (It's always fun to try to "transmit" an object you really loathe once in a while.)
8. *Write down in your journal a detailed description of what you sent and then what your partner received.*

   Then switch and have senders be receivers and vice versa.
9. *Write down in your journal a detailed description of what your partner sent and you received.*

   Keep these descriptions safe to keep a record of your accuracy through the year.

Powering Up

The following are the rudimentaries of the program I developed for maximizing your use of intention. For the complete program, consult *The Intention Experiment*.

Although the power of intention is such that any sort of focus may have some effect, the scientific evidence suggests that you will be a more effective "intender" if you believe in the process, learn how to focus, quiet your mind, connect with the object of your intention, visualize the outcome, mentally rehearse and let go, trusting the process.

1. Choose Your Intention Space

A number of scientific studies suggest that your intention works faster and better if you use the same intention space each time. Choose a place to carry out your intentions that feels comfortable, a place where you and your group can sit quietly and meditate.

2. Focus Your Mind

Powering up involves developing the ability to attend with peak intensity, moment by moment. One of the surest ways to develop this is to practice maintaining your concentration in the present and focusing on your five senses while involved in everyday activities. You can practice turning off the constant inner chatter of your mind and concentrate on your sensory experiences while engaging in everyday activities like eating your cornflakes, waiting in line, putting on your coat, or even walking to work. One good means of harnessing your mind to the present is to "come into your body" and check in with your individual senses. In time you will be able to attend in your intention group with peak intensity.

Sit in a comfortable position in a chair. Breathe slowly and rhythmically in through the nose and out through the mouth (slowly blow all the air out), so that your in-breath is the same length as your out-breath. Allow the belly to relax so that it slightly protrudes, then pull it back slowly as if you were trying to get it to touch your back. This will ensure that you are breathing through your diaphragm.

Repeat this every fifteen seconds, but ensure that you are not overexerting or straining. Carry on for three minutes and then keep observing it. Work up to five or ten minutes. Begin to focus your attention just on the breath, then slowly take an inventory of your five senses. What does the present moment look like? Sound like? Taste like? Feel like? Smell like? Practice this repeatedly.

3. Make a Connection

Touch, or even focus, on the heart or compassionate feelings for the other is a powerful means of causing a "hyperbrain" between people. If you're intending for a member of your group, first form an empathetic connection with him or her by spending a few moments exchanging some personal information about each of you, or even an object or photograph. Hold his or her hands for a deeper connection or spend a little time meditating together.

4. Be Compassionate

Use the following methods to encourage a sense of universal compassion during your Power of Eight group:

- Focus your attention to your heart, as though you are sending light to it. Observe the light spreading from your heart to the rest of your body. Send a loving thought to yourself, such as "May I be well and free from suffering."
- On the out-breath, imagine a white light radiating outward from your heart. As you do, think: "I appreciate the kindnesses and love of all living creatures. May all others be well." As Buddhists recommend, first think of all those you love, then your good friends. Move on to acquaintances and finally to those people you actively dislike. For each stage, think: "May they be well and free from suffering."

5. Tell the Universe Exactly What You Want

Make your intentions highly specific and directed—the more detailed, the better. If you are trying to heal the fourth finger of your left hand, specify that finger and, if possible, the problem with it.

State your entire intention, and include what it is you would like to change, to whom, when, and where. Use a reporter's checklist to ensure you have covered every specific: who, what, when, where, why, and how. Draw a picture of it, or create a collage from photos or magazine pictures. Place this somewhere that you can look at often.

Don't be shy about announcing your intention openly to your group and allow them to hold it for you while you hold intentions for them. *Make a vow, out loud to your group, that you will do everything in your power to make this intention a reality.* As many of my Intention Masterclass members say, having to make a public commitment "to the universe" through the group forces them to keep working harder on their intentions and follow through.

If you are trying to improve your career, don't just say something like "I want money to come in effortlessly." That's far too general.

- If you need more people to sign up for a program of yours, specify how many.
- If something isn't working for you in your work life, work out what it is. The people? The marketing of something? Your role? Tease out the issue and focus on intending for that to change.
- If you want a particular job, write down a full and detailed job spec.

- If your income isn't steady, ask for a very specific job or situation that is likely to offer you a steady flow of money.
- If you want to meet a special other, describe him or her in detail. Draw a mental and physical picture.

6. Mentally Rehearse

The best way to send an intention is to visualize the outcome you desire with all your five senses. You can create mental pictures for anything: a new house, a new job, a new relationship, a healthier body, or a healthier mind. Imagine yourself (or the target of your intention) engaging in whatever new aspect of life you wish to create.

Visualizations don't have to be strictly visual. Some of us are kinesthetic, and have an acute sense of feel; others are auditory, and think in sounds. Your mental rehearsal will depend on which senses are most developed in your brain.

7. Believe in the Process

Don't allow your rational mind to tell you that the intentions won't work. Keep firmly fixed in your mind the desired outcome and do not allow yourself to think of failure. In some studies of intention, the power of belief enabled people to carry out extreme acts.

8. Time It Right

The evidence shows that intentions work better on days when you feel happy and well in every way. It's not always possible to wait; sometimes you need the intentions to make you feel better. But if you have the choice—wait until you are on top of your game.

9. Move Aside

In your meditative state within your group, relax your sense of self and allow yourself to merge with the target of your intention. After framing your intention, state it clearly and then let it go. Don't think of the outcome. This power does not originate with you—you are just the vehicle for it.

In summary:

- Enter your intention space.
- Power up through meditation.
- Move into peak focus through mindful awareness of the present.
- Get onto the same wavelength by focusing on compassion and making a meaningful connection.
- State your intention and make it specific.
- Mentally rehearse every moment of it with all your senses.
- Visualize, in vivid detail, your intention as established fact.
- Time it right—choose days when you feel happy and well.
- Move aside—surrender to the power of the universe and let go of the outcome.

The Power of Eight Experiments

Assemble a group of your friends who are interested in trying out some group intention exercises. Create an intention space where you will meet each time. Select a group target for your community.

Message in a bottle

Ask one of your members to fill a jar with plain water, "send" an object into the jar via a ten-minute intention meditation (during which time they just focus on the name of the object and imagine it with their five senses). Write the name of the word on a piece of paper, fold it over so the word isn't visible, and wrap it round the jar, securing it with a rubber band.

Have the intender hold the jar and show it to the group (if you are on a web platform like Google Hangouts or Skype). If in a Facebook group, take a photo of the jar and upload to your group page.

The other group members should focus on the jar and try to intuit the word in the jar.

Sending loving thoughts to plants

Try your own Germination Intention Experiments to see if you can make plants grow faster and healthier through the power of intention.

1. Purchase two sets of seeds.
2. Plant both sets.
3. Send loving intention to one set of the seeds to grow a certain specific number of inches by a specific date.
4. After two weeks, measure the outcome on both. See which batch of seedlings have grown higher.

Ask one member of the group to be responsible for buying the seeds, planting the seeds and measuring them. They can even upload the seeds and seedlings on Facebook or your Hangouts group. Choose a time to send intention to the seeds as a group. Have your designated person measure the plants in two weeks to see what has happened.

Purifying water

The easiest way to demonstrate any shift in purification is to measure a change in pH. The lower any pH measure below 7, which is neutral, the more acidic something is and the higher the pH above 7, the more alkaline.

Here's how to try it out in your group:

1. Nominate someone to run the Intention Experiment.
2. Ask them to buy some pH strips from a local drugstore. Have them take two glasses of tap water from the same source, and label them A and B. Nominate one to be target glass for the group and the other, the control.
3. Have them take pH measurements of both glasses of water.
4. Photograph the target glass and upload it to your Google Hangouts group or Facebook page.
5. At a designated time, ask all members of the group to send intention to raise the pH of the water by 1 full pH. Imagine the water as a clear mountain stream.
6. Wait a few minutes until after the intention is finished. Then take another pH measurement of the water. See if it has shifted at all. (Don't worry if it doesn't look like it has—the strips aren't as sensitive as the scientific equipment we've used.)

Send in any other experiments you are doing to: www.lynnemctaggart.com.

Afterword

Handing the Power of Eight to the World

For a decade, I've been the nervous sentinel of the Power of Eight, as though it were a sacred amulet to be carefully protected from the wrong hands. Over the years, when various people considered capitalizing on it after witnessing its miraculous effects, I'd discouraged them from doing so. I wanted these group effects to be used for the right reasons, and not for cashing in on miracles. I also wanted to see how we could scale up the process to use it to heal the world in a meaningful way. But mostly, before unleashing it en masse, I wanted to understand more about what it was that I was witnessing, while trying to work out how on earth I'd managed to embroil myself in delivering it to the world.

By the summer of 2017, I felt ready to begin having groups run themselves in a carefully protected way. My first opportunity to do so arose in August 2017, when I was filming a program about the Power of Eight and the Intention Experiments for Gaia television, which would include a major American Peace Intention Experiment, and I found a willing community in the Mile Hi Church in Denver.

I'd decided to set up a few Power of Eight groups, film them during healing, and then minutely record any changes among the recipients. Reverend Kay Johnson of the Mile Hi Church gamely agreed to collect sixteen volunteers, and we held our first group on Monday, August 28, in a small conference room at a Hyatt hotel in Boulder the day before our three-day shoot, and the second one with the participants seated around a stone labyrinth at Gaia's headquarters outside of Boulder.

At the Hyatt hotel, the group of eight focused on Linda Wilkinson, a woman who'd been diagnosed in 2013 with inoperable stage 4 lung cancer—a cancer that kills more people than breast, colon, and prostate cancer combined. Linda had already beaten the odds—the majority of people with her cancer are dead within a year—largely by following an integrative approach, with very low-dose chemotherapy and a change of diet and supplements, plus a great deal of mental and spiritual work. By the time the Power of Eight group met, the lymphatic system in her chest, which had been filled with cancer, was clearing and her tumor markers, which had been 6.2 when she was first diagnosed, stood at just under three, or "high normal." Within weeks after the Power of Eight group, the markers had dropped to two, within the "low normal" range, with at most two small active spots on her lungs. During the circle, she said, "I could feel a shift and I could feel that people really wanted me to be well. It wasn't just my desire—it was the desire of everyone around me. To me, it was a feeling of being lifted and assisted, holding each other up in a higher vibration of a way to be."

Natalie Lancaster put herself forward as the other receiver at the Hyatt, asking that the intention be that her family's "business expand in all directions and be prosperous in all ways." Their computer support business had largely been built on just two major clients, and she and her husband, Mark, had just heard that one of their two clients was about to take their business elsewhere. Thirteen years before, as victims of a Ponzi scheme, she and Mark had lost more than half a million dollars and had to file for bankruptcy. Natalie had to take a job in real estate "flipping houses" in order to survive. Now, it seemed, their financial state was again precarious, and at the time of the intention, Natalie was haunted and depressed about the prospect of having money troubles once again.

During the Power of Eight group, Natalie kept visualizing both her and Mark opening their car doors and money spilling out. Afterward, Natalie's depression and fear seemed to lift. She decided to go back into real estate as a stopgap measure, and worked for two companies, but then other opportunities seemed to spontaneously open up. The Lancaster business received a steady stream of smaller clients, and within a few months, she and her husband learned that the major client they'd been afraid they'd lose had decided to stay. Aside from diversifying their client list, Natalie, who paints as a hobby, received an offer to work one-on-one with an expert in her style of painting who has applauded her talent. She's trained in Qigong, and hopes to hold art and Qigong workshops herself. Mark, meanwhile, is interested in starting a sideline business training program, funded by a small sum he received a few months after his father's death. In Natalie's mind, the Power of Eight group helped her and her husband trust in self-empowerment and the possibility of prosperity, and because of that a load of opportunities have arisen "out of the blue, a process that keeps unfolding even up to yesterday."

One of the receivers around the labyrinth at Gaia was Connie Wiggins, who asked for help healing a misaligned temporomandibular joint in her jaw. Connie had pain all over her head and back, with a good deal of neck pain. On a scale of one to ten, she said, "I live in the seven to eight range."

During her Power of Eight group, Connie found her mind quieting and calming, something she ordinarily found difficult to achieve during meditation, an extraordinary sense of peace entering her body, and an intense outpouring of love. Immediately afterward, Connie's pain level lowered to one or two out of ten. Two days later, Connie reported that her pain was nearly gone.

But the strangest thing was that after the circle had finished, she continued to feel the energy throughout her body. "It wasn't just in that ten minutes," she said a few hours afterward. "I'm aware of a lot more peace and it seems to be continuing. Something happened in that group."

In both of the two circles, many of the senders also underwent major transformations.

Wes Chapman, a burly sixty-five-year-old with a big-hearted laugh, was among the eight at the Hyatt hotel. Although he'd suffered from depression for years, he didn't want to put himself forward as the receiver as he felt that Linda's and Natalie's needs were greater.

But Wes had ample reason to be chosen as the receiver. In 1971, he'd been in his sophomore year at Colorado State University studying physical science, with hopes of either going to medical school or carrying out medical research, when the US government abolished draft deferments for college men. Wes was reclassified 1-A, and as his number in the draft lottery was lower than those who were drafted the year before, he realized that he would likely be sent to Vietnam that year. Wes knew what he was about to face. The Pentagon Papers, the classified Department of Defense report detailing a litany of government deceptions about the course of the war and the scale of American casualties, had just been published in the *New York Times* and *Washington Post*. To avoid being drafted into the Army infantry and facing ground combat, Wes enlisted in the Air Force, but he scored so highly on his vocational tests that the Air Force assigned him to combat intelligence training and he ended up serving in a combat zone during the final chaotic years of the war.

The experience proved so traumatic that Wes returned home deeply depressed. "I had a lot of dreams and hopes, but somehow the prospect of being drafted just made me give that all up. I went into survival mode, just to make it through wartime," he said. "And then my life after that seemed to continue in patterns of darkness or suffering and even death to a surprising extent, and an exercise in survival for years."

The one bright spot in his life was his marriage in 1999 to his second wife, but she developed a fast-growing cancer and he lost her just seven years later. Her death left him devastated, both emotionally and financially. The cost of her treatment had been so exorbitant that he'd had to give up their house. He worked as a long-distance trucker for six years, until her medical bills were paid off, sleeping in the truck's bunk every night, and relying on truck stops across America for washing and laundry.

After retiring a few years ago, that pattern of depression continued, he said, and by the time I met Wes, he'd descended into many years of "what's the use" despondency. The smallest chore—even making breakfast for himself—had become a difficult ordeal. The only activity he enjoyed was attending the Mile Hi Church and his fifteen-year-old contemplative morning rituals. "Frankly, my meditation practice is what's kept me going."

During the Power of Eight group session, Wes felt a strong energy running through the group's hands and a change in the energy of the room. "I felt a very special sort of atmosphere was being created, high vibration, very pure, a little piece of heaven, call it what you will, but it seemed to me there was something special in that room."

After a night's sleep, Wes woke up feeling profoundly different. "The first thing I felt was this energy—I could do what I need[ed] to do, and do it happily. The second thing was amazingly heightened sensory awareness, as well as the emotions that go with those sensations. I had a cup of herbal tea, and it knocked my socks off. And I went outside, and I noticed that the flowers and trees were just so beautiful. It was like all my senses were multiplied by five or ten."

He also discovered that he was suddenly and uncharacteristically social, both to people he knew and to total strangers. "I had definitely gotten into the habit of avoiding people. Now when I see someone coming, I automatically feel enjoyment. And these good feelings come so quickly and so easily."

Of all his spiritual practices, Wes had experienced nothing like it before. "I've done a lot of work over the years, but this was the most phenomenal, the most powerful, and the most direct."

The second night after his Power of Eight group, Wes had an extraordinarily lucid dream—more like a vision, it was so real. "I found myself back on the campus where I was drafted, and there was a chapel there I'd always liked and walked past every day, the only sacred space on that campus. My nineteen-year-old self was waiting for me, and I felt the most powerful emotions of love and connection and joy. All of his high hopes, all of his beautiful dreams, all of his optimism, all of his youthful energy, somehow, I'd left all that behind, in that trauma. But to meet him again and he's been waiting there so patiently for me, it was like I got it back. My nineteen-year-old self has been salvaged and plugged back in."

Wes's younger self also came with a message—not so much with words, but with some sort of intuitive understanding—that he shouldn't feel sad about his lost years and that there was still time to enjoy his life. "Now, with my experience and wisdom and strength and toughness, and his optimism and hope and joy, it's like, hey, I get to start life all over again, at sixty-five."

In the months that followed, Wes's shift carried on and the chronic depression he'd lived with for so many years lifted. He began writing every day and embarked on an intense exercise program. "My inner nineteen-year-old seems very determined that 'we' need to get in better physical shape." Wes found himself engaging in ninety-minute power walks and lengthy bouts of weight lifting. The stiffness and soreness he felt was at most 10 percent of the pain he'd felt during such workouts before. It was as though his nineteen-year-old energy was affecting his metabolism and enabling his body to condition more rapidly.

He also learned how to recognize when he was slipping back into bad mental habits, and how to get back in touch with that surge of "sweet energy" he had felt the first day after his experience in the Power of Eight group. "It never fails. In a minute or less it gives me an express elevator ride from the basement of my self-created misery *up* to the joy, peace, and brilliant light of the top floor, where I intend to dwell from now on."

Melinda Jacobs, another one of the senders, is an experienced meditator, intuitive, and an empath, but for her, the group experience was profoundly different from ordinary meditation. She'd only experienced that sense of expansion and connectedness—in which she'd lost a sense of her own individuality and become part of a larger entity—just a few times before in her life. "I would almost say the word 'bliss.' I definitely didn't want to come out of it." In addition to an extraordinarily intense palpable energy, behind every person in the circle, she said, "there was this light, a ball of light—a light being of sorts—as if they were the magnifiers and they were behind us and their energy was coming through us. It was very profound to feel us contained within another circle of light behind us."

Two days later, Melinda reported that she'd experienced major transformations in all her family relationships. Prior to coming to the Power of Eight group, she'd had an extremely difficult conversation with a family member that had left her so upset that it had "rocked her world," but after returning home, she'd received a very different phone call from the same person. "The person would normally have been very defensive, shut down, and walled up. That was completely gone. They were talking and engaging, and I was feeling calm, [despite] how upsetting the topic had been."

Later that afternoon, she got a call from another family member, who asked for help, a request completely outside the usual family dynamic, and then another call from a third family member, who admitted that he was worried about his health and not having a job. Both calls were stunningly unusual displays of vulnerability and emotion. "Just to have that phone call, 'I need help'—I've never heard those words before from that second family member—and for family member number three to identify with an emotion, that's never happened in my fifty-one years, so it was really just amazing.

"For every one of them, it was like I was the pebble from the experience being dropped into the water, and the ripple just threw my entire family outside their normal dynamic. It was just shocking."

In the months that followed, Melinda found that the greatest change had occurred in herself, and she found the courage to move away from Denver. "The Power of Eight group had a huge impact on my ability to finally break away from very deep family patterns. And I have reestablished a relationship that is so much healthier and loving than I've ever experienced before! We have come back in a much stronger, healthier version of ourselves. We are now living together again in a new city, a new home, and a new life. I don't even have words for the magnitude of change that has occurred in my life since participating in the group experience!"

Six weeks later, I returned to Mile Hi Church to help the church form Power of Eight circles among an audience of five hundred. Once again, I was witness to dozens of people undergoing instant, extraordinary healings.

Sixty-three-year-old Sande Cournoyer had been a lifelong athlete, but one of her knees was now shot. "When I walk I can feel when my knee is going to pop out, which it does, and then I fall. I'm in a lot of pain." She'd scheduled knee replacement surgery for a month's time.

During the intention circle, tears rolled down her cheeks as she felt pressure around her knee on both sides, "as if somebody with big mitts was holding my leg. It was warm, not hot. I've never felt that before. It went down my whole leg, going down to my ankle. When we opened our eyes, we looked at our hands, and everyone's hands were vibrating, with lots of tears."

When the intention was finished and I asked the audience to share any experiences, Sande was the first to put up her hand. "Look," she said, "I can bend my knees." She leaned over and squatted down. "I could not do that before." Three days later, she reported, "When I walk I don't get that feeling that the knee will pop out. I can go upstairs without pain." Since the Power of Eight group experience, Sande has no longer needed her brace, and when the improvement persisted she was able to cancel her surgery.

Beverly Sparks, a massage therapist, hadn't even planned to come to the session because she'd been recovering from a car accident. "My crunched ribs were uncomfortable," she said. "My shoulder was singing a high-pitched, exquisite pain and wouldn't stay in place without support." Nevertheless, a church group member was particularly persuasive, and Beverly arrived, holding her shoulder in place. During the Power of Eight circle, she was the receiver. "I felt all the tension of the car wreck fall out of my left side. Supported by my group mates, bubbling, Kundalini-type energy poured out onto the floor through my arm. My shoulder slid back to normal position, my ribcage released, and I took a full breath for the first time in the six weeks since the accident." Nearly two months later, Beverly reported back that her shoulder had remained in place "almost exclusively pain-free."

And there were many others that night: Joan, who'd had two mini-strokes and could no longer focus her eyes, was able to see normally; a fellow with bursitis could suddenly raise his arm all the way, as normal; another woman who'd arrived with a migraine said that after the circle it had completely cleared; one of the senders who'd arrived with a cane walked away from the event no longer needing it. During her group circle, Faith Cole felt the ache in her back immediately begin to dissipate—to her amazement. "I'm a real skeptic," she said, "and yet it worked."

From September 30 through October 5, 2017, I ran a daily American Peace Intention Experiment via a filmed broadcast on Gaia television. I'd decided to choose an area of high violence and settled on a section around Natural Bridge Avenue in northern St. Louis, Missouri, designated by the *Wall Street Journal* as the most dangerous street in America, with murder rates three times that of Chicago and fifteen times that of New York City. The bridge traverses the entire city east to west, and one of the worst neighborhoods along that stretch is Fairground, bordered by West Florissant Avenue to the north and Natural Bridge Avenue to the south. Fairground consistently achieves a D or F in its report card for violent crime, with one of highest violent crime rates per capita in the country, and equally poor marks for schools, housing, and employment. According to one real estate rating system, the overall crime rate in Fairground is 321 percent higher than the national average, with one in every nine residents falling victim to personal or property crime. Nearly half of the formerly grand Victorian houses, now mainly converted to apartments for rent, stand vacant year-round, their rear walls collapsing, their interiors ripped bare of fixtures. Fairground's major claim to fame is a string of negative superlatives: a higher vacancy rate, a higher rate of child poverty, lower car ownership, and lower per capita income than 99 percent of the rest of the country. Natural Bridge Avenue, so named for a natural limestone bridge that enabled horse-drawn wagons to cross over the now dried-up Rock Branch Creek, makes an occasional appearance in rap songs—"meet me on the bridge"—connoting an invitation to settle the score with a gun. It's the kind of place where thirty-five-year-old father of six Brandon Ellington could head out to the drugstore for a minor errand and never come back, shot dead by his own protégé after he discovered that Brandon was carrying a large tax rebate in cash.

In our Intention Experiment, we decided to focus on violent crime alone—including rape, robbery, aggravated assault, and murder—and as with the other Peace Intention Experiments, I quantified our intention: to lower violent crime by at least 10 percent.

Dr. Jessica Utts, the professor of statistics who analyzed the results of our first Peace Intention Experiment, examined four sets of crime data in St. Louis from September 2014 through March 2018: monthly violent and property crimes in Fairground and St. Louis as a whole. She then used all this the data from the months and years preceding our Intention Experiment to forecast what should happen in the months immediately following our intervention if nothing changed and compared them to what actually did happen in the six months after our experiment. She found that actual crimes were higher than forecast for property and violent crime in the city as a whole, as was property crime in Fairground. Only Fairground's violent crime—the target of our intention—had a lower incidence than forecast. When Dr. Utts mapped the historical trends of crime in the area, she found that although violent crime in Fairground had been steadily increasing, that general trend reversed immediately after our Intention Experiment.

According to the city's official police statistics and the US Census Bureau, as compiled by the *St. Louis Post-Dispatch*, all crime in Fairground was up by 7.14 percent in the six months from October 2017 to March 2018, compared to October to March of the preceding year, but the increase in crime was solely due to an increase in property crime, because violent crime—again, the target of our intention—had in fact fallen over the same period by 43 percent. The year before, during the same six-month period, the police had recorded forty-four violent crimes; that figure had plunged to twenty-five after our experiment. Fairground was one of the only neighborhoods along Natural Bridge Avenue that enjoyed this large decrease in violent crime.

Did we do this? I have no idea, although with each Peace Intention Experiment, the evidence that group consciousness has the power to affect violence gets a bit more compelling. And once more, a survey of my participants showed a profound mirror effect on all who took part.

"I have had an amazing turnaround in my relationship with my teenage stepdaughter—like we are both interacting with love as opposed to fear—miraculous! It's been bad for seven years!!!"

"A true miracle happened. It is healing [the relationship] between all of us, my mother and her daughters, and even a smoother energy in the relationships as a whole in the family."

"My cousin, who had been estranged from my life, contacted me and opened up communication."

"My father apologized to me for disowning me one and a half years ago."

"I was having a very difficult time with a new boss. . . . The week after I did the Intention Experiment, things suddenly smoothed out unexpectedly. It's a night-and-day difference at work now."

"My husband has transformed! He is kind, considerate, and treats me with respect—what a long-awaited turnaround! Even my friends have noticed he is different!"

"My estranged son has suddenly more time for me, calls me and comes to see me. For twenty years we have been practically strangers, but our relationship has drastically improved."

"A better relationship with my wife who did not participate in the experiment herself. . . . *She* seems to me to be more open and less likely to become defensive."

Three-quarters reported extraordinary changes: more in love with their lives, more tolerant of people not like them, a greater desire to work for peace.

"I have since participated in Project Uplift, where I went by van to various gathering spots of homeless people to help serve hot meals and hand out toiletries, clothing, blankets, and pet food. I've never done anything like that before."

"I began a new job with a healing arts clinic. . . . I had not up until now started carving out a career in the healing arts arena."

"It charged my spiritual battery."

An even greater effect occurred among the participants of the Middle Eastern Peace Intention Experiment I ran a month later, on November 9 of that year. I'd been in touch with Tsipi Raz, an Israeli documentary filmmaker, about carrying out an Intention Experiment for Jerusalem, and by some amazing, fortuitous synchronicity, the date we'd proposed for the experiment coincided with when in conjunction Dr. Salah Al-Rashed was running a Middle Eastern summit remotely from a studio he'd acquired in the United Kingdom. He and Tsipi began meeting with me via Skype to plan an Intention Experiment for peace with both Arab and Israeli participants, made possible because of Salah's ingenious technological setup.

In a small backwater warehouse in southern England, Salah has created SmartsWay Studio, with the technological capacity to create a two-way communication flow between a facilitator sitting in the studio and audiences in nine different locations anywhere in the world. For this particular summit, he'd arranged for cameras and monitors to be placed in hotel conference rooms, each full of Arabs, in various cities in Saudi Arabia, Kuwait, Abu Dhabi, Oman, Bahrain, Jordan, and Tunisia. With his permission, I arranged with Tsipi for the ninth camera and monitor to be placed in Gerard Bechar Center in Jerusalem, where hundreds of Israeli Jews would be gathered together for a special one-day peace event, culminating in our experiment. All nine rooms were displayed on my screen, and my feed was visible to all nine audiences so they could observe me running the experiment. The entire experiment and its aftermath were also beamed live to other participants from all over the world via my YouTube channel.

All of us came together to send intention to the Old City of Jerusalem. We chose the site as a symbolic gesture of peace because it is the spiritual heart of Judaism, Christianity, and Islam, a place that belongs to the entire world without preference for a single faith, and we'd targeted the Damascus Gate because it has suffered increasing violence after new security measures were put in place.

Through the ingenuity of Salah's technical team, the technology was interactive; I could call on people from any one of the nine audiences, at which point their screen would be shown to me and the other eight locations, with each group able to speak to me and to the people in the other eight windows.

"You have no idea how revolutionary this is," one Saudi woman remarked to me. "We have never seen Israelis. We've been taught to believe that they have horns coming out of their heads."

Both Arabs and Jews were crying and laughing as they recognized the common humanity in each other. A woman from Abu Dhabi said she had seen visions of Israelis dancing with Palestinians during the experiment. A woman from Jerusalem described her visions of Israeli soldiers hugging Arabs, and another envisioned a wedding with an Arab man dancing with a Jewish woman. "I love you, I love you so much," said Fatima from Jeddah in Saudi Arabia to the Jews in Jerusalem, blowing them kisses. "Your God is my God."

"We love you, sister," called out people from the Israeli audience.

"It's so overwhelming, the possibility of being connected with you, our sisters in Amman and Damascus and in Iran and all over," said Lily, speaking for several Israeli women's peace groups. "We are hundreds of thousands of women here, saying 'Enough!' It's a time of compassion. It's a time of healing. Thank you, dearest sisters, we are one."

Salah and I looked at each other in near disbelief. This was truly history unfolding before our eyes.

News of the experiment broke immediately all across the Middle East. That night, Salah showed me the Twitter feed of a prominent parliamentary member in Kuwait, who acknowledged the experiment as a wonderful mechanism for promoting peaceful relations. As *Time Out Israel* wrote, "While peace in the Middle East is neither black nor white, it's the small efforts like this meditation event that slowly start to make a difference."

And then there was the reaction of our thousands of participants on YouTube. They came from all over the United States, from Brooklyn to Los Angeles, and the United Kingdom, most of Europe, the Arab countries, and Jerusalem, but also more exotic, far-flung places: Australia, Brazil, Japan, Thailand, Hungary, Finland, Columbia, South Africa.

"While I was waiting on YouTube and watching all the people in the chat box send messages of love, peace, healing, etc., I immediately could feel the energy," wrote one participant. "And to look at the number of people that were just there. And to see where everyone was coming in from across the planet, that was overwhelming just in itself, and I was already crying before we even got to the live feed."

"I first started working in that region in Saudi Arabia in 1975, so to hear Arabs from SA send love to Jews in Israel just blew me away," wrote another. "I am also aware of the restrictive cultures that . . . exist in the Middle East and thus the courage . . . exhibited by the Arabs who publicly took part in this intention—that, too, was deeply moving to me."

"My entire perception of the people of the Middle East really changed. . . . We mostly only hear about the violence there, so it truly brought it home to see faces and hear voices of people who care deeply. . . . And this gave me so much hope, it brought it into reality, beyond just my own hopes and prayers. I felt a part of something much larger, much more powerful than myself."

"When you get this opportunity to not only feel like a collective particle but to know on some level that you are one particle of trillions, it changes your perspective on yourself and others."

Did we lower violence? To be honest, I didn't ask the question. As I only had the opportunity to run this for a single day, rather than our usual protocol of running the experiment repeatedly for at least six days, we decided not to examine any changes in violence. Nevertheless, Dr. Roger Nelson examined the random event generator output for his Global Consciousness Project, for this and the St. Louis experiment, and discovered that the machines had reacted strongly during the time of the experiment, in just the direction they had in every earlier Peace Intention Experiment, another tiny shift toward a more coherent world consciousness.

I remain hopeful, as always, about our capacity to decrease the number of dead and injured, but the target of our intention is no longer the point. The point is not the action but its reaction—a ripple effect of peace in the hearts of the participants that could eventually extend out to the entire world. Perhaps these experiments carry one simple truism: To solve a seemingly intractable political situation, the fastest and most effective way forward in a war zone may not be through the military, politics, diplomacy, or even economic initiatives. All you may need are people coming together as a group and praying as one.

The Power of Eight groups at Mile Hi carry on meeting—and healing. Ken no longer experiences pain in his knee; Diane, who'd felt pain in her breast for several weeks and was scheduled for a mammogram, recently got the all clear; Reverend Kay herself, who leads the group and is never the receiver, nonetheless healed her stubborn irritable bowel. Once I left Denver, it was time to release this tool to the world. For those who don't have seven friends nearby for their own Power of Eight group, I have set up a place on my website (www.lynnemctaggart.com/forum) where people can advertise for or join virtual groups in their time zone. I am working to bring the Power of Eight into churches, organizations, communities, healing practices of all persuasions, and businesses. I continue to ask our community to send Intentions of the Week every Sunday. Although many of the nominees we choose have very late-term diseases, a number experience improvement—even healing. Anni from Austria, who had advanced breast cancer, just discovered her cancer was gone; Liana Grace Palermo's intense pelvic and leg pain after breaking her coccyx is 90 percent better; Steven Pacheco, who'd been unable to stand for more than a few minutes at a time after multiple injuries to his back and knees, is back to work; Robert Withers's mother, Mary, who'd become breathless after her pacemaker was installed, is now breathing better, her energy back to normal.

And even the senders are experiencing healing. Christina Wolff, an artist who'd found it difficult to paint, had been faithfully carrying out our Intentions of the Week most Sundays, each time intending for those we had selected, despite the crippling arthritis developing in her own hands. "My fingers started to be crooked, I was in agony, they were swollen, so were my toes, when I moved my fingers they were grating, bone to bone." A few months ago, she wrote me to say that her arthritis was gone, thanks to her involvement in the weekly intentions. "YES CURED," she wrote. Now she's back to painting and cooking—everything she couldn't do before.

I didn't consciously choose this path. I discovered this process largely by accident, but most of the time it feels as though I am a kind of amanuensis, faithfully taking my dictation from the universe. I carry on measuring, documenting, refining our understanding of the nuances and outer boundaries of this ancient miracle, offering the scientific proofs that are palatable to the modern Western mind. Now that we have the capability to connect audiences around the globe and allow them to interact, Salah and I are planning larger Peace Experiments involving every continent on the planet. Every so often, a reader reminds me why I do this. "You know, you are doing more than 'research,'" wrote one participant after the Middle East experiment, "you are breaking open people's hearts." I remain the Power of Eight's gatekeeper, charged with protecting its integrity, but the treasure inside ultimately belongs to the world. It belongs to you and me. Healing in a group is a natural part of your birthright, a capacity that you were born with and that was there all along for you to make use of, not unlike finding out that all you need to do is click your heels together to be instantly transported home. Don't squander this gift. Find your group of eight and discover it for yourself.

Take Part in Life-Transforming Events with Lynne McTaggart

Lynne McTaggart carries out regular live Intention Experiments involving tens of thousands of participants from many different countries, sending specific intentions for peace and an end to discord or suffering in specific areas around the world.

As *The Power of Eight* describes, these experiments not only heal the target, but heal the participants as well.

These events are always free, and all you need to do is register for them. To get involved in the next life-changing event, first join Lynne's community at www.lynnemctaggart.com. You will then receive advance notice of the next Intention Experiment and full instructions about how to take part. You can also read full reports about past Intention Experiments on her website.

As part of Lynne's community, you'll also be able to join in with the Intentions of the Week, where the community sends a collective intention to individuals with health challenges, and you can hear about the activities and outcomes of Power of Eight groups around the world.

You'll also find out more about Lynne's Power of Eight courses, Intention Masterclasses, workshops, retreats, and downloadable materials.

If you don't have seven friends to meet with regularly, you can also set up your own virtual Power of Eight group by visiting www.lynnemctaggart.com/forum.

www.lynnemctaggart.com

Acknowledgments

This project would never have gotten off the ground without the willingness of a number of prestigious scientists and tens of thousands of my readers to participate in what would seem to be, to the ordinary observer, a patently crazy idea.

All the stories in this book have been carefully documented from surveys or personal accounts by the participants themselves. In the majority of cases these are real names, except when the individuals themselves, for one reason or another, requested that I use a pseudonym.

I am beyond grateful for the willingness of the participants in my Intention Experiments, workshops, teleseminars, and e-community to allow me to share their stories, most particularly Todd Voss, for agreeing to take part in our first human Intention Experiment. Special blessings are reserved for my Intention Masterclass 2015 for volunteering to be the first well-monitored group of guinea pigs, and to all the students at Life University who came forward to take part in our experiments.

I am forever indebted to Dr. Gary Schwartz and his then laboratory assistant Mark Boccuzzi, to Dr. Melinda O'Connor, Dr. Konstantin Korotkov, the late Dr. Rustum Roy, Dr. Jessica Utts, Dr. Robert Jahn and Brenda Dunne, Dr. Roger Nelson, the late Dr. Fritz-Albert Popp, Dr. Manju Rao, Dr. Tania Slawecki, Dr. Stephanie Sullivan, and Dr. Jeffrey Fannin for generously donating their time and energy to set up and carry out scientific experiments on my behalf. The late Dr. Masaru Emoto holds a special place in this project for his audacious suggestion that we try to take these experiments out of the laboratory and into the field.

Among the many angels who facilitated this project I include Dr. Guy Riekeman, for putting his university's psychology department at my disposal, and Jim Walsh, for a generous donation that helped to get our Peace Intention Experiment off the ground.

Thanks are also due to Tani Dhamija, Joy Banerjee, and Sameer Mehta for donating their time, facilities, and equipment to create our first Peace Experiment websites, and to Dr. Paul Drouin and Alexi Drouin at Quantum University and Jirka Rysavy of Gaia TV for providing the broadcasting facilities for several Intention Experiments. Dr. Salah Al-Rashed, Dr. Kumar Rupesinghe, Tadzik Greenberg, and Carsten Jacobsen were instrumental in helping me carry out the two major Peace Intention Experiments, Caitlin and Kyle Whelan helped with some of the statistical analysis, and Anya Hubbard assisted in selecting several of our experimental targets.

I am grateful to Dr. Jeff Levin, Klaas-Jan Bakker, Dr. Larry Dossey, and Professor Timothy Darvill for advising me on the effects of prayer and healing in various traditions, and to Dr. Andrew Newberg, Dr. Fred Travis, and Dr. Mario Beauregard for educating me on the neuroscience of meditation and various states of enlightenment and ecstasy.

Thanks are especially due to the editorial and legal assistance of Leslie Meredith, Peter Borland, Daniella Wexler, Mark LaFlaur, and Elisa Rivlin at Atria, all of whom improved the book in countless ways. I am, as always, indebted to my agent Russell Galen, whose dedication to me and to the welfare of this project has been nothing short of astonishing, and to Drew Gerber and all at Wasabi for helping me to tell the world about the Power of Eight.

Several teams at my company WDDTY Publishing Ltd have my gratitude for their involvement in the development of this project. And, finally, this book truly would not have been written without my husband, Bryan Hubbard, and his loving support and guidance, but most of all his gentle insistence that this was a story that needed to be told.

Notes

CHAPTER TWO

14 Over the course of more than two and a half million trials: R. G. Jahn et al., "Correlations of Random Binary Sequences with Pre-stated Operator Intention: A Review of a 12-Year Program," *Journal of Scientific Exploration* 11, no. 3 (1997): 345–67; Dean Radin and Roger Nelson, "Evidence for Consciousness-Related Anomalies in Random Physical Systems," *Foundations of Physics* 19, no. 12 (1989): 1499–1514; Lynne McTaggart, *The Field: The Quest for the Secret Force of the Universe* (New York: HarperCollins, 2002): 116–17.

14 The late William Braud, a psychologist and the research director: William Braud and Marilyn Schlitz, "A Methodology for the Objective Study of Transpersonal Imagery," *Journal of Scientific Exploration* 3, no. 1 (1989): 43–63; W. Braud et al., "Further Studies of Autonomic Detection of Remote Staring: Replication, New Control Procedures and Personality Correlates," *Journal of Parapsychology* 57 (1993): 391–409; M. Schlitz and S. La Berge, "Autonomic Detection of Remote Observation: Two Conceptual Replications," in D. Bierman, ed., *Proceedings of Presented Papers: 37th Annual Parapsychological Association Convention* (Fairhaven, MA: Parapsychological Association, 1994): 465–78.

14 During the height of the AIDS epidemic in the 1980s: F. Sicher et al., "A Randomized Double-Blind Study of the Effect of Distant Healing in a Population with Advanced AIDS: Report of a Small Scale Study," *Western*

Journal of Medicine 168, no. 6 (1998): 356–63. For a full description of the studies, see McTaggart, *The Field*, 181–96.

15 The Transcendental Meditation organization, founded by the late Maharishi Mahesh Yogi: M. C. Dillbeck et al., "The Transcendental Meditation Program and Crime Rate Change in a Sample of 48 Cities," *Journal of Crime and Justice* 4 (1981): 25–45.

17 His batch of mediums turned out to have an accuracy rate of 83 percent: G. Schwartz et al., "Accuracy and Replicability of Anomalous After-Death Communication Across Highly Skilled Mediums," *Journal of the Society for Psychical Research* 65 (2001): 1–25.

18 Popp gave his discovery the ponderous title of "biophoton emissions": For a full description of F. Popp's earlier work, see McTaggart, *The Field*, 39.

19 For our trial run, we aimed to replicate the pilot study: For a complete recounting of that first experiment, see Lynne McTaggart, *The Intention Experiment* (New York: Free Press, 2007), 177.

CHAPTER THREE

32 And this time, a few scientists had navigated a path of sorts before us: B. R. Grad, "A Telekinetic Effect on Plant Growth," *International Journal of Parapsychology*, 5 (1963): 117–33; B. R. Grad, "A Telekinetic Effect on Plant Growth II. Experiments Involving Treating of Saline in Stopped Bottles,"*International Journal of Parapsychology* 6 (1964): 473–98; S. M. Roney-Dougal and J. Solfvin, "Field Study of Enhancement Effect on Lettuce Seeds: Their Germination Rate, Growth and Health," *Journal of the Society for Psychical Research* 66 (2002): 129–43; S. M. Roney-Dougal and J. Solfvin. "Field Study of an Enhancement Effect on Lettuce Seeds—Replication Study,"*Journal of Parapsychology* 67, no. 2 (2003): 279–98.

37 The quality and rhythms of emissions significantly changed: E. P. A. Van Wijk and R. Van Wijk, "The Development of a Bio-Sensor for the State of Consciousness in a Human Intentional Healing Ritual," *Journal of International Society of Life Information Science* 20, no. 2 (2002): 694–702.

37 It was the Intention Experiment's first attempt: G. E. Schwartz et al., "Effects of Distant Group Intention on the Growth of Seedlings," Emerging Paradigms at the Frontiers of Consciousness and UFO Research, Society of Scientific Exploration 27th Annual Meeting, June 25–28, 2008, Boulder, CO.

37 Once they have connected: Non-locality was considered to be proven by Alain Aspect et al.'s experiments in Paris in 1982. See A. Aspect et al., "Experimental Tests of Bell's Inequalities Using Time-Varying Analyzers," *Physical Review*

Letters 49 (1982): 1804–7; and A. Aspect, "Bell's Inequality Test: More Ideal Than Ever," *Nature* 398 (1999): 189–90.

38 A few studies with crystals and algae have hinted: See Lynne McTaggart, *The Bond: Connecting Through the Space Between Us* (New York: Free Press, 2011), chapter 1.

CHAPTER FOUR

40 Kirlian made big claims for this light: S. D. Kirlian and V. K. Kirlian, "Photography and Visual Observation by Means of High-Frequency Currents," *Journal of Scientific and Applied Photography* 6 (1964): 397–403.

41 He'd written five books on the subject: Korotkov's most important works on the subject were *Human Energy Field: Study with GDV Bioelectrography* (Paramus, NJ: Backbone Publishing Company, 2002) and *Aura and Consciousness—New Stage of Scientific Understanding* (St. Petersburg: St. Petersburg division of Russian Ministry of Culture, State Publishing Unit "Kultura," 1999).

41 By 2007, the GDV device was widely used as a general diagnosis tool: L. W. Konikiewicz and L. C. Griff, *Bioelectrography—A New Method for Detecting Cancer and Body Physiology* (Harrisburg, PA: Leonard's Associates Press, 1982); G. Rein, "Corona Discharge Photography of Human Breast Tumour Biopsies," *Acupuncture & Electro-Therapeutics Research* 10 (1985): 305–308; K. Korotkov, "Stress Diagnosis and Monitoring with New Computerized 'Crown-TV' Device," *Journal of Pathophysiology* 5 (1998): 227; K. Korotkov et al., "Assessing Biophysical Energy Transfer Mechanisms in Living Systems: The Basis of Life Processes," *Journal of Alternative and Complementary Medicine* 10, no. 1 (2004): 49–57; P. Bundzen et al., "New Technology of the Athletes' Psycho-Physical Readiness Evaluation Based on the Gas-Discharge Visualisation Method in Comparison with Battery of Tests," "SIS-99" Proceedings, International Congress, St. Petersburg, Russia, 1999: 19–22; P. V. Bundzen et al., "Psychophysiological Correlates of Athletic Success in Athletes Training for the Olympics," *Human Physiology* 31, no. 3 (2005): 316–23; K. Korotkov et al., "Assessing Biophysical Energy Transfer Mechanisms in Living Systems: The Basis of Life Processes," *Journal of Alternative and Complementary Medicine* 10, no. 1 (2004): 49–57.

41 Outside Russia, thousands of medical practitioners were using his machines: Clair A. Francomano, MD, Wayne B. Jonas, MD, and Ronald A. Chez, Proceedings: Measuring the Human Energy Field: State of the Science, The Gerontology Research Center, National Institute of Aging, National Institutes of Health, Baltimore, MD, April 17–18, 2002.

41 While Korotkov enjoys the notoriety he has achieved with these practical applications: S. Kolmakow et al., "Gas Discharge Visualization Technique and Spectrophotometry in Detection of Field Efffects," Mechanisms of Adaptive Behavior, *Abstracts of International Symposium*, St. Petersburg (1999): 79. Also, multiple interviews with K. Korotkov dating from March 2006.

42 Korotkov wrote a book about his discoveries: (Please see Korotkov, *Aura and Consciousness*, in first note for page 41 above.)

43 Two Italian physicists at the Milan Institute for Nuclear Physics: E. Del Giudice, G. Preparata, and G. Vitiello, "Water as a Free Electric Dipole Laser," *Physical Review Letters* 61, no. 9 (1988): 1085–88.

43 As Russian scientists have observed: L. P. Semikhina and V. F. Kiselev, "Effect of Weak Magnetic Fields on the Properties of Water and Ice," *Soviet Physics Journal* 31, no. 5, (1988): 351–54, trans. from Zavedenii, *Fizika* no. 5 (1988): 13–17; S. Sasaki et al., "Changes of Water Conductivity Induced by Non-Inductive Coil," *Society for Mind-Body Science* 1 (1992): 23.

43 and more recently, Luc Montagnier, the Nobel laureate: C. Cardella et al., "Permanent Changes in the Physico-Chemical Properties of Water Following Exposure to Resonant Circuits," *Journal of Scientific Exploration* 15, no. 4 (2001): 501–18; L. Montagnier et al. "DNA Waves and Water," *Journal of Physics: Conference Series* 306, no. 1 (2011): 012007; L. Montagnier et al., "Electromagnetic Signals Are Produced by Aqueous Nanostructures Derived from Bacterial DNA Sequences," *Interdisciplinary Sciences: Computational Life Sciences*. 1 (2009): 81–90. Also, I. Bono et al., "Emergence of the Coherent Structure of Liquid Water," *Water* 4 (2012): 510–32.

43 His equipment has been able to distinguish the infinitesimal differences: K. Korotkov, "Aura and Consciousness," *Journal of Alternative and Complementary Medicine* 9, no. 1 (2003): 25–37; K. Korotkov et al., "The Research of the Time Dynamics of the Gas Discharge Around Drops of Liquid," *Journal of Applied Physics* 95, no. 7 (2004): 3334–38.

44 Dr. Emoto had become well-known for a series of informal experiments: Masaru Emoto, *The Hidden Messages in Water* (New York: Atria Books, 2005).

44 As outrageous as his work seemed: D. I. Radin et al., "Effect of Distant Intention on Water Crystal Formation," *Explore* 2, no. 5 (September/October 2006): 408–11; D. I. Radin et al., "Water Crystal Replication Study," *Journal of Scientific Exploration* 22, no. 4 (2008): 481–93.

48 They even challenged certain Newtonian laws: The full title of Newton's major treatise is *Philosophiae Naturalis Principia Mathematica*, a name that offers

a nod to its philosophical implications, although it is typically referred to reverentially as *The Principia.*

CHAPTER FIVE

54 Many researchers still take the lead from Stonehenge's first archaeologist: W. Stukeley, *Stonehenge, a Temple Restor'd to the British Druids.* London: Printed for W. Innys and R. Manby, 1740. 12, as cited in http://www.voicesfromthedawn.com/stonehenge/

54 "The whole purpose of Stonehenge is that it was a prehistoric Lourdes": H. Wilson, "The Healing Stones: Why Was Stonehenge Built?", February 17, 2011. BBC History website, http://www.bbc.co.uk/history/ancient/british_prehistory/healing_stones.shtml; also author's interview with Timothy Carvill, January 26, 2016.

55 Arthurian legend of the Round Table: Manly P. Hall, *The Secret Teachings of All Ages: An Encyclopedia Outline of Masonic, Hermetic, Qabbalistic and Rosicrucian Symbolical Philosophy* (New York: TarcherPerigee, 2003), 584–91.

56 A number of other practices offered some parallels with the psychic internet: Interview with Jan-Klaas Bakker, October 7, 2016.

56 Many books of the Bible, such as Acts, Ezra, and Jonah: Specifically, those sections Acts 13:1–23, Ezra 8:22–23, and Jonah 3:6–10, as pointed out in *Seven Benefits of Praying Together* by Dr. Jonathan Oloyede, http://www.methodist.org.uk/media/646259/dd-explore-devotion-sevenbenefitsofprayingtogether-0912.pdf. For material on St. Teresa de Ávila, see Mario Beauregard and Denyse O'Leary, *The Spiritual Brain: A Neuroscientist's Case for the Existence of the Soul* (London: HarperOne, 2007): 284.

56 When studying uses of group prayer in Christianity: C. Spurgeon, "The Church on Its Knees: Unleashing the Power of United Prayer," as reproduced on http://www.keepbelieving.com/sermon/the-church-on-its-knees-unleashing-the-power-of-united-prayer.

57 The Authorized King James version of the Bible: Here and elsewhere in this book, I am referring to R. Carroll and S. Prickett, eds. *The Bible: Authorized King James Version with Apocrypha*, Oxford University Press, 2008.

57 Elsewhere it has been translated to mean "with one mind and with one passion": C. Spurgeon, "The Church on Its Knees"; for all Greek translations of *homothumadon* I am indebted to http://biblehub.com/greek/3661.htm.

58 The nineteenth-century American Presbyterian pastor and biblical scholar Albert Barnes: Here and elsewhere in this chapter I am indebted to http://

www.studylight.org/commentaries for such a thorough round-up of the most notable commentary on the Acts and *homothumadon*; A. Barnes, "Commentary on Acts 1:14," *Notes on the New Testament* (Grand Rapids, MI: Baker Books, 2001). http://www.studylight.org/commentaries/acc/acts-1.html.1870.

58 Praying in this manner may have brought the apostles closer: R. Jamieson, A. R. Fausset, and D. Brown, "Commentary on Acts 1:14." *Commentary Critical and Explanatory on the Whole Bible.* Volume 3, CreateSpace Independent Publishing Platform (February 16, 2017); http://www.studylight.org/commentaries/fju/acts-1.html. 1871–78.

58 Seventeenth-century English nonconformist theologian Matthew Poole: M. Poole, "Commentary on Acts 1:14." M. Poole, *Annotations upon The Holy Bible: Wherein The Sacred Text Is Inserted, and Various Readings Annexed, Together With the Parallel Scriptures*, Arkose Press, 2015. http://www.studylight.org/commentaries/mpc/acts-1.html. 1685.

58 British clergyman, dean of Canterbury, and archdeacon: "Commentary on Acts 1:14." "Cambridge Greek Testament for Schools and Colleges." http://www.studylight.org/commentaries/cgt/acts-1.html.

59 More recently, Peter Pett, a retired Baptist minister and university lecturer: P. Pett, Commentary on Acts 1:4, *Peter Pett's Commentary on the Bible*, http://www.studylight.org/commentaries/pet/acts-1.html.

59 Presbyterian minister and former US Senate chaplain Lloyd Ogilvie: Lloyd John Ogilvie, *Drumbeat of Love* (Waco, TX: Word Books, 1976), 19–20.

59 As congregations we cannot be empowered until we are of one mind and heart: Ogilvie, *Drumbeat*, 20.

59 If so, one word that appears is *kahda*: Matthew Black, *An Aramaic Approach to the Gospels and Acts* (Peabody, MA: Hendrickson Publishing, 1967), 10.

59 In Luke (9:1), Jesus gave his apostles "power and authority: Luke (9:1–2) and Matthew (10:1) and (10:8).

60 In Acts, a "multitude out of the cities": Acts of the Apostles (5:16).

60 In his commentary, the eighteenth-century British Methodist biblical scholar Adam Clarke also noted about *homothumadon*: A. Clark, "Commentary on Acts 2:4," *The Adam Clarke Commentary*. http://www.studylight.org/commentaries/acc/acts-2.html.

60 I thought about the words of Clarke: http://www.studylight.org/commentaries/acc/acts-2.html.

60 I looked up the biblical Greek word *ekklésia*: Carroll and. Prickett, *Bible,* op. cit. For definitions of *ekklésia*: http://biblehub.com/greek/1577.htm.

CHAPTER SIX

64 When the TM organization rolled out the study: M. C. Dillbeck et al., "The Transcendental Meditation Program and Crime Rate Change in a Sample of 48 Cities," *Journal of Crime and Justice* 4 (1981): 25–45.

64 They'd also been able to show that: J. Hagelin et al., "Effects of Group Practice of the Transcendental Meditation Program on Preventing Violent Crime in Washington, DC: Results of the National Demonstration Project, June–July 1993," *Social Indicators Research* 47, no. 2 (1999): 153–201.

64 The organization had even experimented with attempts to lower conflict in the Middle East: W. Orme-Johnson et al., "International Peace Project in the Middle East: The Effects of the Maharishi Technology of the Unified Field," *Journal of Conflict Resolution* 32 (1988): 776–812.

69 Despite these first inroads, in 2008 there was still no end in sight: "Sri Lanka's Return to War: Limiting the Damage," Asia Report, no. 146 (February 20, 2008): http://www.refworld.org/pdfid/47bc2e5c2.pdf.

CHAPTER SEVEN

76 On their side, the Tamil Tiger rebels repulsed an army advance: Paul Tighe, "Sri Lanka Battles Tamil Rebels in Land, Air and Sea Attacks," Bloomberg.com (September 18, 2008): http://ourlanka.com/srilankanews/sri-lanka-battles-tamil-rebels-in-land-air-and-sea-attacks-bloomberg.com.htm.

78 After all, the Sri Lankan government's army had increased in size: *Sarath Kumara*, "Fighting intensifies as Sri Lankan army advances on LTTE stronghold," September 29, 2008, World Socialist Website, http://www.wsws.org/en/articles/2008/09/sril-s29.html.

CHAPTER EIGHT

83 It is the moment when, as Saint Teresa de Ávila wrote: T. Butler-Bowdon, *50 Spiritual Classics: Timeless Wisdom from 50 Great Books of Inner Discovery, Enlightenment and Purpose* (London: Nicolas Brealey, 2005), 255; M. Beauregard and D. O'Leary, *The Spiritual Brain: A Neuroscientist's Case for the Existence of the Soul* (London: HarperOne, 2007), 191.

83 *The Course in Miracles* refers to it: *The Course in Miracles,* London: Arksana, 1985; 280.

84 At the end of his life, psychologist Abraham Maslow turned his attention: A. H. Maslow, *Religions, Values, and Peak-Experiences* (Stellar Books, 2014), 33.

85 Most mystical experiences include a profoundly physical component: Dr. Andrew Newberg and Mark Robert Waldman, *How Enlightenment Changes Your Brain: The New Science of Transformation* (London: Hay House, 2016), 40.

87 He'd had the sensation of being part of an enormous force field: For a full description of Edgar Mitchell's epiphany, see Lynne McTaggart, *The Field: The Quest for the Secret Force of the Universe* (New York: HarperCollins, 2002), 6–7.

88 In *The Varieties of Religious Experience*, William James described: William James, *The Varieties of Religious Experience* (New York: New American Library, 1958), 67, quoted in Andrew M. Greeley, *Ecstasy: A Way of Knowing* (Englewood Cliffs, NJ: Prentice Hall, 1974), 8–9.

90 It was Greeley's view that anyone undergoing this state: A. Greeley, *The Sociology of the Paranormal* (Beverly Hills, CA: Sage Publications, 1975), as quoted in J. Levin and L. Steele, "The Transcendent Experience: Conceptual, Theoretical, and Epidemiologic Perspectives," *Explore* 1, no. 2 (2005): 89–101.

CHAPTER NINE

93 But Enlightenment is something else: Andrew Newberg, MD, and Mark Robert Waldman, *How Enlightenment Changes Your Brain: The New Science of Transformation* (London: Hay House, 2016), 43.

93 Newberg discovered that feelings of calm, unity, and transcendence: Andrew Newberg, *Why God Won't Go Away* (New York: Ballantine, 2001), 103.

93 "At the moment they experienced a sense of oneness or loss of self": Newberg and Waldman, *Enlightenment*, 53.

93 The person," wrote Newberg later: Newberg and Waldman, *Enlightenment*, 52.

93 Ultimately, the meditators and praying nuns experienced: Newberg, *Why God Won't Go Away*, 118–19.

94 "Normally, there's a constant dialogue going on": Newberg and Waldman, *Enlightenment*, 94.

94 enlarging it "until it becomes perceived by the mind": Newberg, *Why God Won't Go Away*, 121–22.

94 He distances himself from the strict materialists: Newberg, *Why God Won't Go Away*, 126–27.

95 Could the altered state have been set off by the music I'd been playing: K. Livingston, "Religious Practice, Brain, and Belief," *Journal of Cognition and Culture* 5 (2005): 1–2.

96 Church members describe the experience as the words: Andrew Newberg, MD, and Mark Robert Waldman, *Why We Believe What We Believe: Uncovering Our Biological Need for Meaning, Spirituality, and Truth* (New York: Free Press, 2006), 195.

96 As with his earlier studies, Newberg discovered a sudden drop in frontal lobe activity: *Why God Won't Go Away*, 200–205.

97 In her classic book *Mysticism* Evelyn Underhill writes: Evelyn Underhill, *Mysticism* (New York: E. P. Dutton, 1912).

97 From a neurological point of view, as Newberg describes it: Newberg and Waldman, *Enlightenment*, 91.

CHAPTER TEN

104 Dr. Andrew Newberg once carried out a survey of more than two thousand people: Andrew Newberg, MD, and Mark Robert Waldman, *How Enlightenment Changes Your Brain: The New Science of Transformation* (London: Hay House, 2016), 91.

104 As Abraham Maslow wrote, this is "the way the world looks": As quoted in J. Levin and L. Steele, "The Transcendent Experience: Conceptual, Theoretical, and Epidemiologic Perspectives," *Explore* 1, no. 2 (2005): 89–101.

104 Andrew Greeley discovered that people who had undergone a mystical experience: Andrew M. Greeley, *Ecstasy: A Way of Knowing* (Englewood Cliffs, NJ: Prentice Hall, 1974).

104 Newberg had also discovered evidence that people who experienced: Newberg, *Why God*, 108; Newberg and Waldman, *Enlightenment*, 64–65.

104 In fact, in one study, terminal cancer patients who had undergone: E. C. Kast, "Attenuation of Anticipation: A Therapeutic Use of Lysergic Acid Diethylamide." *Psychiatry Quarterly* 41 (1967): 646–57.

104 The scientific literature contains many case studies of patients: J. Levin and L. Steele, "The Transcendent Experience."

CHAPTER ELEVEN

107 In the Hindu tradition, the point of yoga (union): J. Levin and L. Steele, "The Transcendent Experience: Conceptual, Theoretical, and Epidemiologic Perspectives," *Explore* 1, no. 2 (2005): 89–101.

108 As noted social commentator Barbara Ehrenreich recounts: Barbara Ehrenreich, *Dancing in the Streets: A History of Collective Joy* (London: Granta Books, 2007).

108 Even modern secular group rituals such as those used at a camp: G. Harris, "Healing in Feminist Wicca," in L. L. Barnes and S. S. Sered, eds, *Religion and Healing in America* (New York: Oxford University Press, 2005), 258–61.

108 Deborah Glik, a professor of Community Health Sciences at the University of South Carolina: D. C. Glik, "Symbolic, Ritual and Social Dynamics of Spiritual Healing," *Social Science & Medicine,* 27 no. 11 (1988): 1197–1206.

108 Ted Kaptchuk, director of the Harvard-wide Program in Placebo Studies: Both quotations from T. J. Kaptchuk, "Placebo Studies and Ritual Theory: A Comparative Analysis of Navajo, Acupuncture and Biomedical Healing," *Philosophical Transactions of the Royal Society B* 366 (2011): 1849–58.

109 These kinds of practices are so powerfully effective, he wrote: Kaptchuk, "Placebo Studies."

109 So removing people from their day-to-day environment: Robbie Davis-Floyd, "Research Paper on Rituals," unpublished, citing: Eugene G. d'Aquili et al., *The Spectrum of Ritual: A Biogenetic Structural Analysis* (New York: Columbia University Press, 1979).

109 Indeed, Kaptchuk claims that placebo effects: Kaptchuk, "Placebo Studies."

110 "Once the individuals are gathered together, a sort of electricity": E. Durkheim, *Les Formes Élémentaires de la Vie Religieuse* (Paris: F. F. Alcan, 1915), as quoted in R. Fischer et al., "The Fire-Walker's High: Affect and Physiological Responses in an Extreme Collective Ritual," *PLoS One* 9, no. 2 (2014): e88358.

110 Durkheim also argued that once an individual experienced this state: As noted by J. Haidt et al., "Hive Psychology, Happiness, and Public Policy," *Journal of Legal Studies* 37, no. S2 (June 2008): S133–S156.

111 The scientists concluded that a key ingredient in the association: S. Tewari et al., "Participation in Mass Gatherings Can Benefit Well-Being: Longitudinal and Control Data from a North Indian Hindu Pilgrimage Event," *PLoS One* 7, no. 10 (2012), DOI: 10.1371/journal.pone.0047291.

111 Native Maoris report getting a "fire walker's high": Fischer et al., "The Fire-Walker's High."

111 Even group events using repetitive sounds like drumming: B. Bittman et al., "Composite Effects of Group Drumming Music Therapy on Modulation of Neuroendocrine-Immune Parameters in Normal Subjects," *Alternative Therapies in Health and Medicine* 7, no. 1 (2001): 38–47.

112 According to Dr. Stanley Krippner, professor of psychology: S. Krippner in Marilyn Schlitz et al., *Consciousness & Healing: Integral Approaches to Mind-Body Medicine* (Atlanta, GA: Elsevier, 2005), 179.

112 Cosmic Consciousness spills over into everyday life: F. Travis, "Transcendental Experiences during Meditation Practice," *Annals of New York Academy of Science: Advances in Meditation Research: Neuroscience and Clinical Applications* 1307 (2014): 1–8; F. Travis, and J. Shaw, "Focused Attention, Open Monitoring and Automatic Self-Transcending: Categories to Organize Meditations from Vedic, Buddhist and Chinese Traditions," *Consciousness and Cognition* 19 (2010): 1110–19.

113 Acting in synchrony, engaging in a common intention statement: P. Reddish et al., "Let's Dance Together: Synchrony, Shared Intentionality and Cooperation," *PLos One* 8, no. 8 (2013), DOI: 10.1371/journal.pone.0071182; S. S. Wiltermuth and C. Heath, "Synchrony and Cooperation," *Psychological Science* 20, no. 1 (2009): 1–5, DOI: 10.1111/j.1467-9280.2008.02253.x.

113 Harvey Whitehouse, a statutory chair of social anthropology at University of Oxford: Q. D. Atkinson and H. Whitehouse, "The Cultural Morphospace of Ritual Form," *Evolution and Human Behavior* 32, no. 1 (2011): 50–62.

113 For instance, in research examining the level of stress: M. P. Aranda, "Relationship Between Religious Involvement and Psychological Well-Being: A Social Justice Perspective," *Health and Social Work* 33, no. 1 (2008): 9–21; M. P. Aranda et al., "The Protective Effect of Neighborhood Composition on Increasing Frailty Among Older Mexican Americans: A Barrio Advantage?" *Journal of Aging and Health* 23, no. 7 (2011): 1189–1217.

114 In his research on many religious faiths, Jeff Levin has discovered: J. Levin, "How Faith Heals: A Theoretical Model," *Explore* 5, no. 2 (2009): 77–96.

115 Many experts on the mystical experience concur: J. Levin, interview with author, September 2, 2015; Beauregard and O'Leary, *The Spiritual Brain*, 291.

115 Maslow says that those undergoing this experience invariably feel: Maslow, *Religion, Values, and Peak-Experiences* (Stellar Classics, 1964); E. Underhill, *Mysticism*.

115 The Indian guru Sri Aurobindo once claimed that "bringing down" the "supermind": S. Aurobindo and Kamaladevi R. Kunkolienker, "From 'Mind' to 'Supermind': A Statement of Aurobindonian Approach," The Paideia Project On-line, Twentieth World Congress of Philosophy (Boston, MA, August 10–15, 1998); https://www.bu.edu/wcp/Papers/Mind/MindKunk.htm.

115 The feeling of perfect integration, symbolized by a giant circle: D. Meintel and G. Mossière, "Reflections on Healing Rituals, Practices and Discourse in Contemporary Religious Groups," *Ethnologies* 33, no. 1 (2011): 19–32.

116 People who can trust others enough to be vulnerable enjoy improvements in immune function: J. W. Pennebaker, "Writing about Emotional Experiences as a Therapeutic Process," *Psychological Science* 8 (1997): 162–66; J. W. Pennebaker and M. E. Francis, "Cognitive, Emotional, and Language Processes in Disclosure," *Cognition and Emotion* 10, no. 6 (1996): 601–26, as referenced in J. Levin, "How Faith Heals: A Theoretical Model."

116 Pennebaker has also studied the social dynamics of opening up: H. Dienstfrey, "Disclosure and Health: An Interview with James W. Pennebaker," *Advances in Mind-Body Medicine* 15, no. 3 (1999): 161–63.

CHAPTER TWELVE

122 A number of laboratory experiments had demonstrated: Carroll B. Nash, "Test of Psychokinetic Control of Bacterial Mutation," *Journal of the American Society for Psychical Research* 78 (1984): 145–52.

122 Water is a chemical anarchist, behaving like no other liquid in nature: E.

Stanley, "Liquid Water: A Very Complex Substance," *Pramana Journal of Physics* 53, no. 1 (1999): 53–83.

122 It is most of what we're made of (humans are about 70 percent water): London South Bank University has a very good rundown of water's anarchic behavior: www.lsbu.ac.uk/water.

123 Rusty and his coauthors had synthesized all current research: R. Roy et al. "The Structure of Liquid Water: Novel Insights from Materials Research: Potential Relevance to Homeopathy," *Materials Research Innovations* 9, no. 4 (2005): 1433-075X.

124 "It is this range of very weak bonds": Email correspondence with R. Roy, spring 2009.

125 Canadian studies had shown that when water used to irrigate plants: B. Grad, "Dimensions in 'Some Biological Effects of the Laying on of Hands' and Their Implications," in H. A. Otto and J. W. Knight (eds.), *Dimension in Wholistic Healing: New Frontiers in the Treatment of the Whole Person* (Chicago: Nelson-Hall, 1979): 199–212.

125 and Russian research demonstrated when healing is sent to a sample of water: L. N. Pyatnitsky and V. A. Fonkin, "Human Consciousness Influence on Water Structure," *Journal of Scientific Exploration* 9, no. 1 (1995): 89.

126 In deciding on this equipment, Rusty had been inspired: L. Zuyin, *Scientific Qigong Exploration* (Malvern, PA: Amber Leaf Press, 1997), as reported in R. Roy et al., "The Structure of Liquid Water: Novel Insights."

CHAPTER THIRTEEN

131 The slowest-growing plant was watered by the vial: B. Grad, "The 'Laying on of Hands': Implications for Psychotherapy, Gentling and the Placebo Effect," *Journal of the Society for Psychical Research* 61, no. 4 (1967): 286–305.

132 A statistical analysis on these numbers reached borderline significance: The statistical analysis on our results was ($p < 0.07$), which reached borderline significance because $p < 0.05$ is the minimum considered statistically significant.

134 There was a precedent for using thoughts to affect pH: W. Tiller et al., *Conscious Acts of Creation: The Emergence of a New Physics* (Walnut Creek, CA: Pavior Publishing, 2001), 175, 216.

CHAPTER FOURTEEN

142 Ten months later, Lissa published her book: L. Wheeler, *Engaging Resilience: Heal the Physical Impact of Emotional Trauma* (Charleston, SC: CreateSpace, 2017).

152 "Most of the deviations are negative," he wrote me: Nelson's Global Consciousness Project provides a full analysis of our results: http://teilhard.global-mind.org/intention.110911-18.html.

CHAPTER FIFTEEN

161 Mohandas Gandhi, who believed that all religions "were as dear as one's close relatives": M. Gandhi, as quoted by http://www.mkgandhi.org/my_religion/01definition_of_religion.htm

161 "We aren't talking about superficial team-building exercises": Ruth Braunstein et al., "The Role of Bridging Cultural Practices in Racially and Socioeconomically Diverse Civic Organizations," *American Sociological Review* 79, no. 4 (August 2014): 705–25.

CHAPTER SEVENTEEN

173 In my book *The Bond*, I write about the discovery by Italian neuroscientist Giacomo Rizzolati: V. Gallese et al., "Action Recognition in the Premotor Cortex," *Brain* 119, no. 2 (1996): 593–609.

174 "A brain region that controls the movement of a violinist's fingers": M. Ricard et al., "Mind of the Meditator," *Scientific American* (November 2014): 39–45.

175 As I discovered when I wrote about this in *The Intention Experiment*: McTaggart, *The Intention Experiment*, 70.

175 At this speed, the brain waves also begin synchronizing throughout the brain: McTaggart, *Intention*, 71.

175 As Davidson's research with monks demonstrated: A. Lutz et al., "Long-Term Meditators Self-Induce High-Amplitude Gamma Synchrony during Mental Practice," *Proceedings of the National Academy of Science* 101, no. 46 (2004): 16369–73.

176 After just a single week of carrying out compassionate meditation: S. Leiberg et al., "Short-Term Compassion Training Increases Prosocial Behavior in a Newly Developed Prosocial Game," *PLoS One* 6, no. 3 (2011), DOI:10.1371/journal.pone.0017798.

177 Repeatedly the brains of his listeners begin to evidence a "resonance response": Interview with M. Beauregard, October 14, 2015.

177 Robert Cialdini, a psychologist formerly at the University of Arizona: R. B. Cialdini et al., "Reinterpreting the Empathy-Altruism Relationship: When One into One Equals Oneness," *Journal of Personality and Social Psychology* 73, no. 93 (1997): 481–94.

CHAPTER EIGHTEEN

181 During the following eight years, he remained symptom-free: George is a pseudonym, but his story was documented by Candy Gunther Brown, an associate professor of Religious Studies at Indiana University and published in her book *Testing Prayer: Science and Healing* (Cambridge, MA: Harvard University Press, 2012).

181 Although his healing had not been as immediate or dramatic: R. Clark, *Changed in a Moment* (Mechanicsburg, PA: Apostolic Network of Global Awakening, 2010).

182 "It seems, then, as if praying is more effective than being prayed for": S. O'Laoire, "An Experimental Study of the Effects of Distant, Intercessory Prayer on Self-Esteem, Anxiety and Depression," *Alternative Therapies on Health and Medicine* 3, no. 6 (1997): 19–53.

182 At the end of this study, he discovered that his volunteers were: K. Pellimer, "Environmental Volunteering and Health Outcomes over a 20-Year Period," *Gerontologist* 50 (2010): 594–602.

183 When faced with each new stressful event, those who'd decided not to lend a hand: M. J. Poulin and E. A. Holman, "Helping Hands, Healthy Body? Oxytocin Receptor Gene and Prosocial Behavior Interact to Buffer the Association between Stress and Physical Health," *Hormones and Behavior* 63, no. 3 (2013): 510–17; M. J. Poulin, "Volunteering Predicts Health among Those Who Value Others: Two National Studies," *Health Psychology* 33, no. 2 (2014): 120–29; M. Poulin et al., "Giving to Others and the Association between Stress and Mortality," *American Journal of Public Health* 103, no. 9 (2013): 1649–55.

183 As Father O'Laoire had discovered, directing your attention: N. Mor and J. Winquist, "Self-Focused Attention and Negative Affect: A Meta-Analysis," *Psychological Bulletin* 128, no. 4 (2002): 638–62.

183 in one study of older Americans, those who gave: W. M. Brown et al., "Altruism Relates to Health in an Ethnically Diverse Sample of Older Adults," *Journal of Gerontology Series B: Psychological Sciences and Social Sciences* 60, no. 3 (May 2005): P143–52.

183 And of all the religious coping behaviors relating to better mental health: H. G. Koenig, "Religious Coping and Health Status in Medically Ill Hospitalized Older Adults," *Journal of Nervous and Mental Disease* 186 (1998): 513–21.

183 Research from Stanford University in California of senior residents: D. Oman et al., "Volunteerism and Mortality among the Community-Dwelling Elderly," *Journal of Health Psychology* 4, no. 3 (1999): 301–16.

184 In fact, those willing to give of their time or money: The Saguaro Seminar: Civic Engagement in America, "Social Capital Community Benchmark Survey," Kennedy School of Government, Harvard University, August 2000; https://www.hks.harvard.edu/saguaro/communitysurvey/docs/survey_instrument.pdf.

185 When people who volunteer are surveyed: A. Luks, "Helper's High: Volunteering Makes People Feel Good, Physically and Emotionally," *Psychology Today* (October 1988).

CHAPTER NINETEEN

191 Dacher Keltner, a psychologist at University of California at Berkeley: Dacher Keltner, *Born to Be Good: The Science of a Meaningful Life* (New York: W. W. Norton, 2009).

192 A closer look at the results revealed something even more fascinating: Keltner, *Born to Be Good*, 232–5.

193 After that simple exercise, as a battery of tests revealed: C. Hutcherson et al., "Loving-Kindness Meditation Increases Social Connectedness," *Emotions* (October 2008): 720–28.

193 David Hamilton, the former medical researcher and author: see David Hamilton, *Why Kindness Is Good for You* (London: Hay House, 2010).

194 However, these cytokines were reduced markedly: M. Clodi et al., "Oxytocin Alleviates the Neuroendocrine and Cytokine Response to Bacterial Endotoxin in Healthy Men," *American Journal of Physiology, Endocrinology and Metabolism* 295, no. 3 (2008): E686–91; also Hamilton, *Kindness*, 90.

194 Oxytocin even plays a key role in turning undifferentiated stem cells: Hamilton, *Kindness*, 108.

194 "When Burners give of themselves and work in healing": F. Gauthier, "Les HeeBeeGeeBee Healers au Festival Burning Man. Trois Récits de Guérison," *Ethnologies* 33, no. 1 (2011): 191–217.

195 If you have to choose one path over the other: B. L. Fredrickson et al., "A Functional Genomic Perspective on Human Well-Being," *Proceedings of the National Academy of Science* 110, no. 33 (2013): 13684–89.

195 Scientists from Boston College discovered this: P. Arnstein et al., "From Chronic Pain Patient to Peer: Benefits and Risks of Volunteering," *Pain Management in Nursing* 3, no. 3 (2002): 94–103.

196 Those who regularly assemble in churches to pray together: H. G. Koenig et al., "The Relationship between Religious Activities and Blood Pressure in Older Adults," *International Journal of Psychiatry in Medicine* 28 (1998): 189–213.

196 enjoy far stronger immune systems: H. G. Koenig et al., "Attendance at Religious Services, Interleukin-6, and Other Biological Parameters of Immune Function in Older Adults," *International Journal of Psychiatry in Medicine* 27 (1997): 233–50.

196 spend far fewer days in the hospital: H. G. Koenig and D. B. Larson, "Use of Hospital Services, Religious Attendance, and Religious Affiliation," *Southern Medical Journal* 91 (1998): 925–32.

196 and are a third less likely to die, even when all other factors are controlled: D. Oman and D. Reed, "Religion and Mortality among the Community-Dwelling Elderly," *American Journal of Public Health* 88 (1998): 1469–75.

196 Scientists believe that those who are now age twenty: R. Hummer et al., "Religious Involvement and U.S. Adult Mortality," *Demography* 36 (1999): 273–85.

196 One study found that those living in a religious kibbutz: J. D. Kark et al., "Does Religious Observance Promote Health? Mortality in Secular vs Religious Kibbutzim in Israel," *American Journal of Public Health* 86 (1996): 341–46.

196 An elevated IL level is a marker of one of the degenerative diseases: H. G. Koenig et al., "Attendance at Religious Services."

197 Although those studies achieved statistical significance, they paled: C. G. Brown et al., "Study of the Therapeutical Effects of Proximal Intercessory Prayer (STEPP) on Auditory and Visual Impairments in Rural Mozambique," *Southern Medical Journal* 103 (2010): 864–69.

198 Hell is not other people. Hell is thinking there are other people: Bryan Hubbard, *The Untrue Story of You* (London: Hay House, 2014).

CHAPTER TWENTY-ONE

220 "When a person chooses to seek Enlightenment through a specific practice": Andrew Newberg, MD, and Mark Robert Waldman, *How Enlightenment Changes Your Brain: The New Science of Transformation* (London: Hay House, 2016), 91.

222 With those subjects, and indeed in most instances of contemplative prayer: Newberg and Waldman, *Enlightenment*, 120.

223 Entire areas of the two brains create synchronized patterns: U. Lindenberger et al., "Brains Swinging in Concert: Cortical Phase Synchronization While Playing Guitar," *BMC Neuroscience* 10, no. 22 (2009): doi: 10.1186/1471-2202-10-22.

223 The same team went on to study guitarists who were improvising: V. Müller et al., "Intra- and Inter-Brain Synchronization during Musical Improvisation on the Guitar," *PLoS One* 8, no. 9 (2013): e73852.

223 Other scientists at the University of Lancaster in the United Kingdom: E. Filho, "The Juggling Paradigm: A Novel Social Neuroscience Approach to Identify Neuropsychophysiological Markers of Team Mental Models," *Frontiers in Psychology* 8 (2015): 799.

224 "Changes in the functional state of the human body": K. Korotkov, "Electrophotonic Analysis of Complex Parameters of the Environment and Psycho-Emotional State of a Person," *WISE Journal* 4, no. 3 (2015): 49–56.

232 French anthropologist Laurent Denizeau: L. Denizeau, "Soirées miracles et guérisons," *Ethnologies* 33, no. 1 (2011): 75–93.

233 In that sense, illness is not only a personal trial: My thanks to Jean Hudon, who works with my French publisher Ariane, for helping with the translation of this passage.

Bibliography

Aknin, L., et al. "Prosocial Spending and Well-Being: Cross-Cultural Evidence for a Psychological Universal." *Journal of Personality and Social Psychology* 104, no. 4 (2013): 635–52.

Allen, K. N., and D. F. Wozniak. "The Integration of Healing Rituals in Group Treatment for Women Survivors of Domestic Violence." *Social Work in Mental Health* 12, no. 12 (2014): 52–68.

Alspach, J. G. "Harnessing the Therapeutic Power of Volunteering." *Critical Care Nurse* 34, no. 6 (2014): 11–14.

Aranda, M. P. "Relationship Between Religious Involvement and Psychological Well-Being: A Social Justice Perspective." *Health and Social Work*, 33, no. 1 (2008): 9–21.

Aranda, M. P., et al., "The Protective Effect of Neighborhood Composition on Increasing Frailty Among Older Mexican Americans: A Barrio Advantage?" *Journal of Aging and Health* 23, no. 7 (2011): 1189–1217.

Arnstein, P., et al. "From Chronic Pain Patient to Peer: Benefits and Risks of Volunteering." *Pain Management in Nursing* 3, no. 3 (2002): 94–103.

Aspect, A. "Bell's Inequality Test: More Ideal Than Ever." *Nature* 398 (1999): 189–90.

Aspect, A., et al. "Experimental Tests of Bell's Inequalities Using Time-Varying Analyzers." *Physical Review Letters* 49 (1982): 1804–7.

Atkinson Q. D., and H. Whitehouse. "The Cultural Morphospace of Ritual Form." *Evolution and Human Behavior* 32, no. 1 (2011): 50–62.

Barnes, Albert. *Acts and Romans.* Vol. 10 of *Notes on the New Testament.* Heritage Edition. Grand Rapids, MI: Baker Books, 2001.

Barnes, Linda L., and Susan Starr Sered, eds. *Religion and Healing in America.* New York: Oxford University Press, 2005.

Beauregard, Mario, and Denyse O'Leary. *The Spiritual Brain: A Neuroscientist's Case for the Existence of the Soul.* London: HarperOne, 2007.

Beischel, J., and G. E. Schwartz. "Anomalous Information Reception by Research Mediums Demonstrated Using a Novel Triple-Blind Protocol." *Explore* 3 (2007): 23–27.

Bhat, R. K., et al. "Correlation of Electrophotonic Imaging Parameters With Fasting Blood Sugar in Normal, Prediabetic, and Diabetic Study Participants." *Journal of Evidence-Based Complementary & Alternative Medicine* (November 6, 2016): 1–8.

Bittman, B., et al. "Composite Effects of Group Drumming Music Therapy on Modulation of Neuroendocrine-Immune Parameters in Normal Subjects." *Alternative Therapies in Health and Medicine* 7, no. 1 (2001): 38–47.

Black, Matthew. *An Aramaic Approach to the Gospels and Acts.* Peabody, MA: Hendrickson Publishing, 1967.

Bono, I., et al. "Emergence of the Coherent Structure of Liquid Water." *Water* 4 (2012): 510–32.

Braud, William, and Marilyn Schlitz. "A Methodology for the Objective Study of Transpersonal Imagery." *Journal of Scientific Exploration* 3, no. 1 (1989): 43–63.

Braud, William, Donna Shafer, and Sperry Andrews. "Further Studies of Autonomic Detection of Remote Staring: Replication, New Control Procedures, and Personality Correlates." *Journal of Parapsychology* 57 (1993): 391–409.

Braunstein, R., et al. "The Role of Bridging Cultural Practices in Racially and Socioeconomically Diverse Civic Organizations." *American Sociological Review* 79, no. 4 (August 2014): 705–25.

Brown, C. G., et al. "Study of the Therapeutical Effects of Proximal Intercessory Prayer (STEPP) on Auditory and Visual Impairments in Rural Mozambique." *Southern Medical Journal* 103 (2010): 864–69.

Brown, Candy Gunther. *Testing Prayer: Science and Healing.* Cambridge, MA: Harvard University Press, 2012.

Brown, W. M., et al. "Altruism Relates to Health in an Ethnically Diverse Sample of Older Adults." *Journal of Gerontology Series B: Psychological Sciences and Social Sciences* 60, no. 3 (2005): P143–52.

Bundzen, P. V., et al. "Psychophysiological Correlates of Athletic Success in Athletes Training for the Olympics." *Human Physiology* 31, no. 3 (2005): 316–23.

Bundzen, P., et al. "New Technology of the Athletes' Psycho-Physical Readiness Evaluation Based on the Gas-Discharge Visualisation Method in Comparison with Battery of Tests," "SIS-99" Proceedings, International Congress, St. Petersburg, Russia, 1999: 19–22.

Burkert, Walter. *Ancient Mystery Cults*. Cambridge, MA: Harvard University Press, 1987.

Burl, Aubrey. *A Guide to the Stone Circles of Britain, Ireland and Brittany.* New Haven, CT: Yale University Press, 2005.

Butler-Bowden, Tom. *50 Spiritual Classics: Timeless Wisdom from 50 Great Books of Inner Discovery, Enlightenment and Purpose* (London: Nicholas Brealey, 2015), 255.

Cardella, C., et al. "Permanent Changes in the Physico-Chemical Properties of Water Following Exposure to Resonant Circuits." *Journal of Scientific Exploration* 15, no. 4 (2001): 501–18.

Chaplin, Martin. "Water Structure and Science." www.lsbu.ac.uk/water.

Cialdini, R. B., et al. "Reinterpreting the Empathy-Altruism Relationship: When One into One Equals Oneness." *Journal of Personality and Social Psychology* 73, no. 93 (1997): 481–94.

Clark, R. *Changed in a Moment*. Mechanicsburg, PA: Apostolic Network of Global Awakening, 2010.

Clodi, M., et al. "Oxytocin Alleviates the Neuroendocrine and Cytokine Response to Bacterial Endotoxin in Healthy Men." *American Journal of Physiology—Endocrinology and Metabolism* 295 (2008): W686–91.

Coruh, B., et al. "Does Religious Activity Improve Health Outcomes? A Critical Review of the Recent Literature." *Explore* 1, no. 3 (2005): 186–91.

Course in Miracles, A. London: Arksana, 1985.

Crisis Group. "Sri Lanka's Return to War: Limiting the Damage." *Asia Report*, no. 146 (February 20, 2008). http://www.refworld.org/pdfid/47bc2e5c2.pdf.

Crone, N. E., et al. "High-Frequency Gamma Oscillations and Human Brain Mapping with Electrocorticography." *Progress in Brain Research* 159 (2006): 275–95.

Davidson, R. J., and A. Harrington. *Visions of Compassion: Western Scientists and Tibetan Buddhists Examine Human Nature*. New York: Oxford University Press, 2002.

Davis-Floyd, R. "Research Paper on Rituals," unpublished, citing Eugene G. d'Aquili et al., *The Spectrum of Ritual: A Biogenetic Structural Analysis*. New York: Columbia University Press, 1979.

Del Giudice, E., et al. "Water as a Free Electric Dipole Laser." *Physical Review Letters* 61 no. 9 (1988): 1085–88.

Denizeau, L. "Soirées miracles et guérisons." *Ethnologies* 33, no. 1 (2011): 75–93.

Diensfrey, H. "Disclosure and Health: An Interview with James W. Pennebaker." *Advances in Mind Body Medicine* 15 (1999): 161–63.

Dillbeck, M. C., et al. "The Transcendental Meditation Program and Crime Rate Change in a Sample of 48 Cities." *Journal of Crime and Justice* 4 (1981): 25–45.

Dossey, L. "Healing Research: What We Know and Don't Know." *Explore* 4, no. 6 (2008): 341–52.

Durkheim, E. *Les Formes Élémentaires de la Vie Religieuse*, Paris: Alcan, 1915.

Easterling, D., and C. G. Foy. "Social Capital Community Benchmark Report" (2006): Winston-Salem, NC: Wake Forest University School of Medicine, Department of Social Sciences & Health Policy.

Ehrenreich, Barbara. *Dancing in the Streets: A History of Collective Joy*. London: Granta Books, 2007.

Eicher, D. J., and S. C. Springer. "Effects of a Prayer Circle on a Moribund Premature Infant." *Alternative Therapies in Health and Medicine* 5, no. 3 (1999): 115–20.

Emoto, Masaru. *The Hidden Messages in Water*. New York: Atria, 2005.

Feather, Robert. *The Secret Initiation of Jesus at Qumran: The Essene Mysteries of John the Baptist*. London: Watkins Publishing, 2006.

Filho, E. "The Juggling Paradigm: A Novel Social Neuroscience Approach to Identify Neuropsychophysiological Markers of Team Mental Models." *Frontiers in Psychology* 8 (2015): 799.

Fischer, R., et al. "The Fire-Walker's High: Affect and Physiological Responses in an Extreme Collective Ritual." *PLoS One* 2, no. 9 (2014): e88355.

Fjorback, L. O., and H. Walach. "Meditation Based Therapies—A Systematic Review and Some Critical Observations." *Religions* 3 (2012): 1–18.

Francomano, Clair A., and Wayne B. Jonas. *Proceedings: Measuring the Human Energy Field: State of the Science*. Edited by Ronald A. Chez. The Gerontology Research Center, National Institute of Aging, National Institutes of Health. Baltimore, MD, April 17–18, 2002.

Fredrickson, B. L., et al. "A Functional Genomic Perspective on Human Well-Being." *Proceedings of the National Academy of Science* 110, no. 33 (2013): 13684–89.

Full, G. E., et al. "Meditation-Induced Changes in Perception: An Interview Study with Expert Meditators (Sotapannas) in Burma." *Mindfulness* 4, no. 1 (2013): 55–63.

Gallese, V., et al. "Action Recognition in the Premotor Cortex," *Brain* 119, no. 2 (1996): 593–609.

Gauthier, F. "Les HeeBeeGeeBee Healers au Festival Burning Man. Trois Récits de Guérison." *Ethnologies* 33, no. 1 (2011): 191–217.

Glik, D. C. "Symbolic, Ritual and Social Dynamics of Spiritual Healing." *Social Science & Medicine* 27, no. 11 (1988): 1197–1206.

———. "Psychosocial Wellness among Spiritual Healing Participants," *Social Science and Medicine* 22, no. 5 (1986): 579–86.

Grad, B. "The 'Laying on of Hands': Implications for Psychotherapy, Gentling and the Placebo Effect." *Journal of the Society for Psychical Research* 61, no. 4 (1967): 286–305.

———. "Dimensions in 'Some Biological Effects of the Laying on of Hands' and Their Implications." In *Dimensions in Wholistic Healing: New Frontiers in the Treatment of the Whole Person*. Edited by Herbert Arthur Otto and James William Knight. Chicago: Nelson-Hall, no. 2, 1979.

———. "A Telekinetic Effect on Plant Growth." *International Journal of Parapsychology* 5 (1963): 117–33.

Greeley, Andrew M. *Ecstasy: A Way of Knowing*. Englewood Cliffs, NJ: Prentice Hall, 1974.

———. *The Sociology of the Paranormal* (Beverly Hills, CA: Sage Publications, 1975), as quoted in J. Levin and L. Steele, "The Transcendent Experience: Conceptual, Theoretical, and Epidemiologic Perspectives." *Explore* 1, no. 2 (2005): 89–101.

Hagelin, J., et al. "Effects of Group Practice of the Transcendental Meditation Program on Preventing Violent Crime in Washington, DC: Results of the National Demonstration Project, June–July 1993." *Social Indicators Research* 47, no. 2 (1999): 153–201.

Haidt, J., et al. "Hive Psychology, Happiness, and Public Policy." *Journal of Legal Studies* 37, no. S2 (2008): S133–56.

Hall, Manly P. *The Secret Teachings of All Ages: An Encyclopedia Outline of Masonic, Hermetic, Qabbalistic and Rosicrucian Symbolical Philosophy*. New York: Tarcher/Penguin, 2003.

Hamilton, David R. *Why Kindness Is Good for You*. London: Hay House, 2010.

Harris, G. "Healing in Feminist Wicca." In *Religion and Healing in America*. Edited by Linda L. Barnes and Susan Starr Sered. New York: Oxford University Press, 2005.

Hart, C., and S. Hong. "Trajectories of Volunteering and Self-Esteem in Later Life: Does Wealth Matter?" *Research on Aging* 35, no. 5 (2013): 571–90.

Harung, H. S., and F. Travis. "Higher Mind-Brain Development in Successful Leaders: Testing a Unified Theory of Performance." *Cognitive Processing* 113, no. 2 (2012): 171–81.

Heaton, D. P., and F. Travis, "Consciousness, Empathy, and the Brain." In *Organizing Through Empathy*. Edited by Kathryn Pavlovich and Keiko Krahnke. Oxford, UK: Routledge, 2013.

Hinds, Arthur. *The Complete Sayings of Jesus*. London: Forgotten Books, republished 2008.

Holy Bible: The Authorized King James Version with Apocrypha. Oxford University Press, 2008.

http://teilhard.global-mind.org/intention.110911-18.html

http://www.studylight.org/commentaries

Hubbard, Bryan. *The Untrue Story of You*. London: Hay House, 2014.

Hummer R., et al. "Religious Involvement and U.S. Adult Mortality," *Demography* 36 (1999): 273–85.

Hutcherson, C., et al. "Loving-Kindness Meditation Increases Social Connectedness." *Emotions* (2008): 720–28.

Hutcherson, C. A., et al. "The Neural Correlates of Social Connection." *Cognitive, Affective, and Behavioral Neuroscience* 15, no. 1 (2015): 1–14.

Jahn, R. G., et al. "Correlations of Random Binary Sequences with Pre-stated Operator Intention: A Review of a 12-Year Program." *Journal of Scientific Exploration* 11, no. 3 (1997): 345–67.

James, William. *The Varieties of Religious Experience: A Study in Human Nature*. New York: Longmans, Green and Co., 1902.

Jamieson, R., et al. *Commentary Critical and Explanatory on the Whole Bible: The New Testament: From Matthew to Second Corinthians*. Volume 3. Charleston, SC: CreateSpace, 2017.

Jantos, M., and H. Kiat. "Prayer as Medicine: How Much Have We Learned?" *Medical Journal of Australia* 186, supplement 10 (2007): S51–53.

Jonas, Wayne B., and Cindy C. Crawford. *Healing Intention and Energy Medicine: Science, Research Methods and Clinical Implications*. London: Churchill Livingston, 2003.

Kaptchuk, T. J. "Placebo Studies and Ritual Theory: A Comparative Analysis of Navajo, Acupuncture and Biomedical Healing." *Philosophical Transactions of the Royal Society B* 366 (2011): 1849–58.

Kark, J. D., et al. "Does Religious Observance Promote Health? Mortality in Secular vs Religious Kibbutzim in Israel." *American Journal of Public Health* 86 (1996): 341–46.

Kast, E. C. "Attenuation of Anticipation: A Therapeutic Use of Lysergic Acid Diethylamide." *Psychiatry Quarterly* 41 (1967): 646–57.

Keltner, Dacher. *Born to Be Good: The Science of a Meaningful Life*. New York: W. W. Norton, 2009.

Kirlian, S. D., and V. K. Kirlian. "Photography and Visual Observation by Means of High-Frequency Currents," *Journal of Scientific and Applied Photography* 6 (1964): 397–403.

Koenig, H. G., et al. "The Relationship between Religious Activities and Blood

Pressure in Older Adults." *International Journal of Psychiatry in Medicine* 28 (1998): 189–213.

Koenig, H. G., et al. "Attendance at Religious Services, Interleukin-6, and Other Biological Parameters of Immune Function in Older Adults." *International Journal of Psychiatry in Medicine* 27 (1997): 233–50.

———. "Religious Coping and Health Status in Medically Ill Hospitalized Older Adults." *Journal of Nervous and Mental Disease* 186 (1998): 513–21.

Koenig, Harold G. *The Healing Power of Faith: How Belief and Prayer Can Help You Triumph Over Disease*. New York: Simon & Schuster, 1999.

———. *Medicine, Religion and Health: Where Science and Spirituality Meet*. Conshohocken, PA: Templeton Press, 2008.

Koenig, H. G., and D. B. Larson. "Use of Hospital Services, Religious Attendance, and Religious Affiliation." *Southern Medical Journal* 91 (1998): 925–32.

Koenig, Harold G., and Saad Al Shohaib. *Health and Well-Being in Islamic Societies: Background, Research, and Applications*. Switzerland: Springer International Publishing, 2014.

Kok, B. E., and B. L. Fredrickson. "Upward Spirals of the Heart: Autonomic Flexibility, as Indexed by Vagal Tone, Reciprocally and Prospectively Predicts Positive Emotions and Social Connectedness." *Biological Psychology* 85, no. 3 (2010): 432–36.

Kolmakow, S., et al. "Gas Discharge Visualisation Technique and Spectrophotometry in Detection of Field Effects." Mechanisms of Adaptive Behavior, Abstracts of International Symposium, St. Petersburg, Russia, 1999.

Konikiewicz, Leonard W., and Leonard C. Griff. *Bioelectrography—A New Method for Detecting Cancer and Body Physiology*. Harrisburg, PA: Leonard's Associates Press, 1982.

Korotkov, Konstantin G. "Stress Diagnosis and Monitoring with New Computerized 'Crown-TV' Device,' *Journal of Pathophysiology* 5 (1998): 227.

———. *Aura and Consciousness—New Stage of Scientific Understanding*. St. Petersburg, Russia: St. Petersburg division of Russian Ministry of Culture, State Publishing Unit "Kultura," 1999.

———. *Human Energy Field: Study with GDV Bioelectrography*. Paramus, NJ: Backbone Publishing, 2002.

———. "New Conceptual Approach to Early Cancer Detection," *Mind and Physical Reality* [in Russian] 3, no. 1 (1998): 51–58; "Aura and Consciousness," *Journal of Alternative and Complementary Medicine* 9, no. 1 (2003): 25–37.

———. "Experimental Research of Human Body Activity after Death," St. Petersburg Federal University SPITMO (St. Petersburg, Russia), 2014.

———. "Science of Measuring Energy Fields: A Revolutionary Technique to Visualize Energy Fields of Humans and Nature." In *Bioelectromagnetic and Subtle Energy Medicine*. Edited by Paul Rosch. Boca Raton, FL: CRC Press, 2015.

———. "Electrophotonic Analysis of Complex Parameters of the Environment and Psycho-Emotional State of a Person." *WISE Journal* 4, no. 3 (2015): 49–56.

Korotkov, K., et al. "Assessing Biophysical Energy Transfer Mechanisms in Living Systems: The Basis of Life Processes." *Journal of Alternative and Complementary Medicine* 10, no. 1 (2004): 49–57.

———. "The Research of the Time Dynamics of the Gas Discharge Around Drops of Liquid." *Journal of Applied Physics* 95 no. 7 (2004): 3334–38.

Koss-Chioino, J. "Spiritual Transformation and Healing: Is Altruism Integral?" In *Altruism in Cross-Cultural Perspective*. Edited by Douglas A. Vakoch. New York: Springer, 2013.

Kuldeep K., et al. "Effect of Yoga Based Techniques on Stress and Health Indices Using Electro Photonic Imaging Technique in Managers." *Journal of Ayurveda and Integrative Medicine* 7 (2016): 119–23.

Kunkolienker, Kamaladevi R. "From 'Mind' to 'Supermind': A Statement of Aurobindonian Approach," The Paideia Project On-line, Twentieth World Congress of Philosophy (Boston, MA, August 10–15, 1998); https://www.bu.edu/wcp/Papers/Mind/MindKunk.htm.

Leiberg, S., et al. "Short-Term Compassion Training Increases Prosocial Behavior in a Newly Developed Prosocial Game." *PLoS One* 6, no. 3 (2011). doi:10.1371/journal.pone.0017798.

Levin, J. "How Prayer Heals: A Theoretical Model." *Alternative Therapies in Health and Medicine* 2, no. 1 (1996): 66–73.

———. "Spiritual Determinants of Health and Healing: An Epidemiologic Perspective on Salutogenic Mechanisms." *Alternative Therapies* 9, no. 6 (2003): 48–57.

———. "Esoteric Healing Traditions: A Conceptual Overview." *Explore* 4, no. 2 (2008): 101–12.

———. "Bioenergy Healing: A Theoretical Model and Case Series." *Explore* 4, no. 3 (2008): 201–9.

———. "How Faith Heals: A Theoretical Model." *Explore* 5, no. 2 (2009): 77–96.

Levin, J. S., and H. Y. Vanderpool. "Is Frequent Religious Attendance Really Conducive to Better Health? Toward an Epidemiology of Religion." *Social Science & Medicine* 24, no. 7 (1987): 589–600.

Levin, J., and L. Steele. "The Transcendent Experience: Conceptual, Theoretical, and Epidemiologic Perspectives." *Explore* 1, no. 2 (2005): 89–101.

———. "On the Epidemiology of 'Mysterious' Phenomena." *Alternative Therapies in Health and Medicine* 7, no. 1 (2001): 64–66.

Lewis, H. Spencer. *The Mystical Life of Jesus*. San Jose, CA: Grand Lodge of the English Language Jurisdiction, AMORC, 2006.

———. *The Secret Doctrines of Jesus*. San Jose, CA: Grand Lodge of the English Language Jurisdiction, AMORC, 2006.

Lindenberger, U., et al. "Brains Swinging in Concert: Cortical Phase Synchronization While Playing Guitar." *BMC Neuroscience* 10, no. 22 (2009): doi: 10.1186/1471–2202–10–22.

Livingston, K. "Religious Practice, Brain, and Belief." *Journal of Cognition and Culture* 5 (2005): 1–2.

Luks, A. "Helper's High: Volunteering Makes People Feel Good, Physically and Emotionally." *Psychology Today* (October 1988).

Lutz, A., et al. "Long-Term Meditators Self-Induce High-Amplitude Gamma Synchrony during Mental Practice." *Proceedings of the National Academy of Science* 101, no. 46 (2004): 16369–73.

———. "Attention Regulation and Monitoring in Meditation." *Trends in Cognitive Sciences* 12, no. 4 (2008): 163–69.

———. "Regulation of the Neural Circuitry of Emotion by Compassion Meditation: Effects of Meditative Expertise." *PLoS One* 3, no. 3 (2008).

Maslow, Abraham H. *Religions, Values, and Peak-Experiences*. Stellar Books, 2014.

McDonough-Means, Sharon I., et al. "Fostering a Healing Presence and Investigating Its Mediators." *Journal of Alternative and Complementary Medicine* 10, supplement 1 (2004): S25–41.

McTaggart, Lynne. *The Field: The Quest for the Secret Force of the Universe*. New York: HarperCollins, 2002.

———. *The Intention Experiment: Using Your Thoughts to Change Your Life and the World*. New York: Free Press, 2007.

———. *The Bond: Connecting Through the Space Between Us*. New York: Free Press, 2011.

Meintel, D., and G. Mossière. "Reflections on Healing Rituals, Practices and Discourse in Contemporary Religious Groups." *Ethnologies* 33, no. 1 (2011): 19–32.

Meyer, Marvin Wayne. *The Ancient Mysteries: A Sourcebook of Sacred Texts*. New York: HarperCollins, 1987.

Moll, J., et al. "Human Fronto-Mesolimbic Networks Guide Decisions about Charitable Donation." *Proceedings of the National Academy of Sciences* 103, no. 42 (2006): 15623–28.

Monroe, Kristen Renwick. *The Heart of Altruism: Perceptions of a Common Humanity*. Princeton, NJ: Princeton University Press, 1996.

Montagnier, L., et al. "DNA Waves and Water." *Journal of Physics: Conference Series* 306, no. 1 (2011): 012007.

———. "Transduction of DNA Information through Water and Electromagnetic Waves." *Electromagnetic Biology and Medicine* 34, no. 2 (2015): 106–12.

———. "Electromagnetic Signals Are Produced by Aqueous Nanostructures Derived from Bacterial DNA Sequences." *Interdisciplinary Sciences: Computational Life Sciences* 1 (2009): 81–90.

Mor, N., and J. Winquist. "Self-Focused Attention and Negative Affect: A Meta-Analysis." *Psychological Bulletin* 128, no. 4 (2002): 638–62.

Müller, V., et al. "Intra- and Inter-Brain Synchronization during Musical Improvisation on the Guitar." *PLoS One* 8, no. 9 (2013): e73852.

Nash, Carroll B. "Test of Psychokinetic Control of Bacterial Mutation." *Journal of the American Society for Psychical Research* 78, no. 2 (1984): 145–52.

Newberg, Andrew. *Why We Believe What We Believe: Uncovering Our Biological Need for Meaning, Spirituality, and Truth*. New York: Free Press, 2006.

———. *Why God Won't Go Away*. New York: Ballantine, 2001.

Newberg, Andrew, and Mark Robert Waldman. *How Enlightenment Changes Your Brain: The New Science of Transformation*. London: Hay House, 2016.

O'Laoire, S. "An Experimental Study of the Effects of Distant, Intercessory Prayer on Self-Esteem, Anxiety and Depression." *Alternative Therapies on Health and Medicine* 3, no. 6 (1997): 19–53.

Ogilvie, Lloyd John. *Drumbeat of Love: The Unlimited Power of the Spirit as Revealed in the Book of Acts*. Waco, TX: Word Books, 1976.

Oliner, Samuel P. *Do Unto Others: How Altruism Inspires True Acts of Courage*. Boulder, CO: Westview Press, 2003.

Oloyede, Jonathan. *Seven Benefits of Praying Together*. London: The Methodist Church in Britain [n.d.]. http://www.methodist.org.uk/media/646259/dd-explore-devotion-sevenbenefitsofprayingtogether-0912.pdf

Oman, D., and D. Reed. "Religion and Mortality among the Community-Dwelling Elderly." *American Journal of Public Health* 88, no. 10 (1998): 1469–75.

Oman, D., et al. "Volunteerism and Mortality among the Community-Dwelling Elderly," *Journal of Health Psychology* 4, no. 3 (1999): 301–16.

Orme-Johnson, W., et al. "International Peace Project in the Middle East: The Effects of the Maharishi Technology of the Unified Field." *Journal of Conflict Resolution* 32 (1988): 776–812.

Panneck, J. "The Ritual Use of Ayahuasca: The Healing Effects of Symbolic and Mythological Participation." http://psypressuk.com/2015/01/16/the-ritual-use-of-ayahuasca-the-healing-effects-of-symbolic-and-mythological-participation. Originally published in *Psychedelic Press Journal*.

Pearce, Joseph Chilton. *The Biology of Transcendence: A Blueprint of the Human Spirit*. Rochester, VT: Park Street Press, 2002.

———. *The Death of Religion and the Rebirth of Spirit: A Return to the Intelligence of the Heart*. Rochester, VT: Park Street Press, 2007.

Pellimer, K., et al. "Environmental Volunteering and Health Outcomes over a 20-Year Period." *Gerontologist* 50 (2010): 594–602.

Pennebaker, J. W. "Writing about Emotional Experiences as a Therapeutic Process." *Psychological Science* 8 (1997): 162–66.

Pennebaker, J. W., and M. E. Francis. "Cognitive, Emotional, and Language Processes in Disclosure." *Cognition and Emotion* 10, no. 6 (1996): 601–26.

Piliavin, J. A., and E. Sieg. "Health Benefits of Volunteering in the Wisconsin Longitudinal Study." *Journal of Health and Social Behavior* 48, no. 4 (2007): 450–64.

Poole, Matthew. *Annotations upon The Holy Bible: Wherein the Sacred Text Is Inserted, and Various Readings Annexed, Together with the Parallel Scriptures*. Welwyn, Hertfordshire, UK: Arkose Press, 2015.

Poulin, M. J. "Volunteering Predicts Health among Those Who Value Others: Two National Studies." *Health Psychology* 33, no. 2 (2014): 120–29.

Poulin, M., et al. "Giving to Others and the Association between Stress and Mortality." *American Journal of Public Health* 103, no. 9 (2013): 1649–55.

Poulin, M. J., and E. A. Holman. "Helping Hands, Healthy Body? Oxytocin Receptor Gene and Prosocial Behavior Interact to Buffer the Association between Stress and Physical Health." *Hormones and Behavior* 63, no. 3 (2013): 510–51.

Pyatnitsky, L. N., and V. A. Fonkin. "Human Consciousness Influence on Water Structure." *Journal of Scientific Exploration* 9, no. 1 (1995): 89–105.

Radin, D., and R. Nelson. "Evidence for Consciousness-Related Anomalies in Random Physical Systems." *Foundations of Physics* 19, no. 12 (1989): 1499–1514.

Radin, D. I., et al. "Double-Blind Test of the Effects of Distant Intention on Water Crystal Formation." *Explore* 2, no. 5 (September/October 2006): 408–11.

———. "Water Crystal Replication Study." *Journal of Scientific Exploration* 22, no. 4 (2008): 481–93.

Reddish, P., et al. "Let's Dance Together: Synchrony, Shared Intentionality and Cooperation." *PLos One* 8, no. 8 (2013). doi: 10.1371/journal.pone.0071182.

Rein, G. "Corona Discharge Photography of Human Breast Tumour Biopsies." *Acupuncture & Electro-Therapeutics Research* 10 (1985): 305–8.

Ricard, Matthieu. *Altruisim: The Power of Compassion to Change Yourself and the World.* London: Atlantic Books, 2013.

Ricard, M., et al. "Mind of the Meditator." *Scientific American* (November 2014): 39–45.

———. "Neuroscience Reveals the Secrets of Meditation's Benefits." *Scientific American* (November 2014).

Roney-Dougal, S. M., and J. Solfvin. "Field Study of an Enhancement Effect on Lettuce Seeds—Replication Study." *Journal of Parapsychology* 67, no. 2 (2003): 279–98.

———. "Field Study of Enhancement Effect on Lettuce Seeds: Their Germination Rate, Growth and Health." *Journal of the Society for Psychical Research* 66 (2002): 129–43.

Roy R. "A Contemporary Materials Science View of the Structure of Water." Symposium on Living Systems/Materials Research, Boston, MA, November 28, 2004.

Roy, R., et al. "The Structure of Liquid Water; Novel Insights from Materials Research; Potential Relevance to Homeopathy." *Material Research Innovations* 9, no. 4 (2005): 1433–075X.

Saguaro Seminar: Civic Engagement in America, "Social Capital Community Benchmark Survey," Kennedy School of Government, Harvard University, August 2000; https://www.hks.harvard.edu/saguaro/communitysurvey/docs/survey_instrument.pdf

Sänger, J., et al. "Directionality in Hyperbrain Networks Discriminates Between Leaders and Followers in Guitar Duets." *Frontiers of Human Neuroscience* 4, no. 7 (2013): 234.

Sasaki, S., et al. "Changes of Water Conductivity Induced by Non-Inductive Coil." *Society for Mind-Body Science* 1 (1992): 23.

Sauer, S., et al. "Spirituality: An Overlooked Predictor of Placebo Effects?" *Philosophical Transactions of the Royal Society of London. Series B, Biological Sciences* 366, no. 1572 (2011): 1838–48.

Schlitz, M., and W. Braud. "Distant Intentionality and Healing: Assessing the Evidence." *Alternative Therapies in Health and Medicine* 3, no. 6 (1997): 62–73.

Schlitz, M., and N. Lewis. "Directed Prayer & Conscious Intention: Demonstrating the Power of Distant Healing." In *Breast Cancer Beyond Convention: The World's Foremost Authorities on Complementary and Alternative Medicine*

Offer Advice on Healing. Edited by Mary Tagliaferri, Isaac Cohen, and Debu Tripathy. New York: Atria Books, 2002.

Schlitz, Marilyn, and Tina Amorok. *Consciousness & Healing: Integral Approaches to Mind-Body Medicine*. Atlanta, GA: Churchill Livingstone/Elsevier, 2005.

Schlitz, M., and S. La Berge. "Autonomic Detection of Remote Observation; Two Conceptual Replications." In *Proceedings of Presented Papers: 37th Annual Parapsychological Association Convention* [Amsterdam]. Edited by D. Bierman. Fairhaven, MA: Parapsychological Association, 1994: 465–78.

Schwartz, G., et al. "Accuracy and Replicability of Anomalous After-Death Communication Across Highly Skilled Mediums." *Journal of the Society for Psychical Research* 65 (2001): 1–25.

———. "Effects of Distant Group Intention on the Growth of Seedlings." Emerging Paradigms at the Frontiers of Consciousness and UFO Research, Society of Scientific Exploration 27th Annual Meeting, June 25–28, 2008, Boulder, CO.

———. "Consciousness, Spirituality and Post-Materialist Science: An Empirical and Experiential Approach." In *The Oxford Handbook of Psychology and Spirituality*. Edited by Lisa J. Miller. New York: Oxford University Press, 2011.

Schwartz, S., and L. Dossey. "Nonlocality, Intention, and Observer Effects in Healing Studies: Laying a Foundation for the Future." *Explore* 5, no. 5 (2010): 295–307.

Semikhina, L. P., and V. F. Kiselev. "Effect of Weak Magnetic Fields on the Properties of Water and Ice." *Soviet Physics Journal*, 31 no 5 (1988): 351–54, trans. from Zabedenii, *Fizika* no. 5 (1988): 13–17.

Seppala, E. "Compassionate Mind, Healthy Body." *Greater Good: The Science of a Meaningful Life* (July 24, 2013).

Sicher, F., et al. "A Randomized Double-Blind Study of the Effect of Distant Healing in a Population with Advanced AIDS: Report of a Small Scale Study." *Western Journal of Medicine* 168, no. 6 (1998): 356–63.

Siegel, Daniel J. *The Mindful Brain: Reflection and Attunement in the Cultivation of Well-Being*. New York: W. W. Norton, 2007.

Simmonds-Moore, Christine, ed. *Exceptional Experience and Health: Essays on Mind, Body and Human Potential*. Jefferson, NC: McFarland & Company, 2012.

Spurgeon, C. "The Church on Its Knees: Unleashing the Power of United Prayer," as reproduced here: http://www.keepbelieving.com/sermon/the-church-on-its-knees-unleashing-the-power-of-united-prayer.

Stanley, E. "Liquid Water: A Very Complex Substance." *Pramana Journal of Physics* 53, no. 1 (1999): 53–83.

Surowiecki, James. *The Wisdom of Crowds: Why the Many Are Smarter Than the Few and How Collective Wisdom Shapes Business, Economies, Societies, and Nations.* London: Abacus, 2004.

Szekely, Edmond Bordeaux, transl. *The Gospel of the Essenes: The Unknown Book of the Essenes / Lost Scrolls of the Essene Brotherhood.* Saffron Walden, UK: C. W. Daniel Co., 2002.

Tewari, S., et al., "Participation in Mass Gatherings Can Benefit Well-Being: Longitudinal and Control Data from a North Indian Hindu Pilgrimage Event." *PLoS One* 7, no. 10 (2012). doi:10.1371/journal.pone.0047291.

Thaut, M. H., et al. "Temporal Entrainment of Cognitive Functions: Musical Mnemonics Induce Brain Plasticity and Oscillatory Synchrony in Neural Networks Underlying Memory." *Annals of the New York Academy of Sciences* 1060 (2005): 243–54.

Thomas, Keith. *Religion and the Decline of Magic.* London: Penguin, 1971.

Tighe, Paul. "Sri Lanka Battles Tamil Rebels in Land, Air and Sea Attacks." Bloomberg.com. September 18, 2008. http://www.bloomberg.com/apps/news?pid=newsarchive&sid=aq0MUmQ01f6o/.

Tiller, W., et al. *Conscious Acts of Creation: The Emergence of a New Physics.* Walnut Creek, CA: Pavior Publishing, 2001.

Tiller, W. A., and W. E. Dibble, Jr. "New Experimental Data Revealing an Unexpected Dimension to Materials Science and Engineering." *Materials Research Innovations* 5 (2001): 21–34.

Travis, F. "Transcendental Experiences during Meditation Practice," *Annals of New York Academy of Science: Advances in Meditation Research: Neuroscience and Clinical Applications* 1307 (2014): 1–8.

Travis, F., and J. Shaw. "Focused Attention, Open Monitoring and Automatic Self-Transcending: Categories to Organize Meditations from Vedic, Buddhist and Chinese Traditions." *Consciousness and Cognition* 19 (2010): 1110–19.

Travis, F., et al. "Moral Development, Executive Functioning, Peak Experiences and Brain Patterns in Professional and Amateur Classical Musicians: Interpreted in Light of a Unified Theory of Performance." *Consciousness and Cognition* 20, no. 4 (2011): 1256–64.

Underhill, Evelyn. *Mysticism: A Study in Nature and Development of Spiritual Consciousness,* New York: E. P. Dutton, 1912.

Van Wijk, E. P. A., and R. Van Wijk. "The Development of a Bio-Sensor for the State of Consciousness in a Human Intentional Healing Ritual." *Journal of the International Society of Life Information Science (ISLIS)* 20, no. 2 (2002): 694–702.

Van Wijk, R., and E. Van Wijk. "The Search for a Biosensor as a Witness of a Human Laying on of Hands Ritual." *Alternative Therapies in Health and Medicine* 9, no. 2 (2003): 48–55.

Van Willigen, M. "Differential Benefits of Volunteering Across the Life Course." *Journal of Gerontology Series B: Psychological Science and Social Science* 55, no. 5 (2000): S308–18.

Williams, D. R., and M. J. Sternthal. "Spirituality, Religion and Health: Evidence and Research Directions." *Medical Journal of Australia*, supplement 10 (2007): S47–50.

Wilson, H., "The Healing Stones: Why Was Stonehenge Built?" February 17, 2011. BBC History. http://www.bbc.co.uk/history/ancient/british_prehistory/healing_stones.shtml

Wiltermuth, S. S., and C. Heath. "Synchrony and Cooperation." *Psychological Science* 20, no. 1 (2009): 1–5.

Wootton, David. *The Invention of Science: A New History of the Scientific Revolution.* London: Penguin, 2015.

Yakovleva, E. G., et al. "Identifying Patients with Colon Neoplasias with Gas Discharge Visualization Technique." *Journal of Alternative and Complementary Medicine* 21, no. 11 (2015): 720–24.

———. "Engineering Approach to Identifying Patients with Colon Tumors on the Basis of Electrophotonic Imaging Technique Data." *Open Biomedical Engineering Journal* 2 (2016): 72–80.

Index

Index

About the Author

Lynne McTaggart, one of the central authorities on the new science and consciousness, is the award-winning author of seven books, including the internationally bestselling *The Intention Experiment* and *The Field*. She is also cofounder and editorial director of *What Doctors Don't Tell You* (www.wddty.com), one of the world's most respected health magazines, and architect of the Intention Experiments, a web-based "global laboratory." A highly sought-after international public speaker, Lynne has also appeared in many documentaries, including *What the Bleep?! Down the Rabbit Hole, I Am*, and *The Abundance Factor*, and is consistently listed as one of the world's one hundred most spiritually influential people. Lynne and her husband, author and WDDTY cofounder Bryan Hubbard, live in London, and have two adult daughters.

www.lynnemctaggart.com